Law of attraction

The big secret

THE LAW OF attraction

The universe IS EVERYTHING, without exceptions: I letterspace matter - energy -. All that exists is a part of the universe and this governed by some laws, like the law of gravity or the law of inertia. All these universal laws can not be right in some occasions and incorrect in another one, they are applicable everywhere and all the times and do not admit exceptions, that leaves the free road to affirm that they are perfect laws to us.

We all are guided for the same laws and this law IS NOT the odd man out, this acting for all here and now, let's want or no. The Law of Attraction suggests us that every person attracts her life everything that in which focuses its thoughts: Answer to the thoughts and you attract his parallel be good or bad.

All that comes to our life stems from a previous thought and all that will come to our life in a future is the result so that we are thinking about the present, it is for this reason that we must put The Law to work of Attraction on our side and waking up to our thoughts.

Without exception, any human being, you have the ability to turn any scarcity into fortress or benefit and to attract health, abundance, harmony, happiness, love, success, money, etc to his life. THE UNIVERSE IS UNLIMITED: THERE IS ENOUGH OF EVERYTHING, FOR ALL.

This Law was known to our forefathers for people but big of history: Platon, Shakespeare, Newton, Leonardo, Galileo, Hugo, Beethoven, Lincoln, Emerson, Edison, Einstein, etc and for the first time all knowledge about this theme joined to create the famous so-called film THE SOUNDBOARD itself in history.

Toward the ending of the year 2004 and after suffering through a series of traumatic events in his life, Rhonda Byrne discovered the secret when her daughter gave him as a present a book's copy: THE SCIENCE TO BECOME RICH, article in 1910 for Wallace D. Watlles. His life was transformed rapidly when putting into practice what you had learned . At that point his great desire and mission was to share in his knowledge with the world, making room for the film THE SECRET.

Million people have seen the film since it was thrown the marketplace in March 2006. Rhonda Byrne also wrote a book once THE SECRET was called just the same, the #1 is New York Times Bestseller and they have printed over 6 million copies. After carrying the secret in the public eye, hundred of people have dedicated themselves to carrying out an investigation themselves, to deepen and putting in practice The Law of the successful Attraction for own benefit.

How it works The Law of Attraction?

The Law of Attraction works through the thoughts. When you think about what you want and you say him IF to that that you want, these activating the pansy and The Law of Attraction this responding and attracting what you want . You can attract people, events, styles of life, riches, success, etc. Also when these telling NOT to something, really these attracting over that that YOU DO NOT WANT . Great part of the time, our thoughts are in automatic pilot and you are attracted to us for the so much whole

Unknowingly; These thoughts generate sentiments
Good or bad. The bad sentiments can be: Anguish, Rabia, Desespero, Culpa, Miedo, Temor etc. The good sentiments can be: Happiness, Admiración, Tranquilidad, Seguridad, Libertad, etc.

Basically what you focus on or you put your attention with thought and sentiment that is these attracting like if you be a magnet. You consider what, you seat that and what is manifested is always the same thing. It is of great importance feeling good, because that sentiment is what you put salt in like a sign toward the universe and you begin to attract over that very thing toward you.

The film The Ley of Attraction relates THE SECRET The Aladdin's Lamp, than you appear to the genius that you tell when rubbing her: - your desires are orders for me comparing Aladino to the human beings and to the genius of the lamp with the universe.

Also it takes our participation to this creative process acting when we perceived a good opportunity of something pertaining to what we desired, when impulse moves us toward any situation in particular, when intuition is pushing us from within, in those precise moments it is necessary to ACT! When there be an inspiring idea, it is necessary to confide in her.

Visualizing is a part of the functioning of The Law of Attraction, when you visualize, you materialize, when you have that image taking shape in your mind, this sighted you will be present at the final result, but you are important to feel than what we want right now this in our hands, taking for granted that we have what we asked for and feeling that joy and that gratitude to have it right now.

With the soundboard leads it mentioned you stop being a secret and you make way for yourself to the different forms, the several methods that The Law of Attraction has to put into practice.

Thoughts:

Thoughts are creation of the mind: The activity of intellect is whatever you are worn-out to intervening existence. Whatever you come from mental nature thought is considered, well be these abstract, rational, creative, artistic, etc.

Each thought generates a frequency and we are broadcasting individuals of thinking waves. These waves of thought carry I get different kinds from frequency: Statuses gamma, beta, alpha, theta and delta and each of these frequencies generates several type of reality and of perception, this mean that we make use of our thoughts in order that our own nature express itself .

We have the capacity to create reality according to our thoughts, because thoughts send a magnetized sign than this carrying back his parallel toward us.

The frequencies of the thoughts can be of two types: Refusals or Positive Prints.

NEGATIVE FREQUENCIES:

- criticize
- hatred
- rancor
- envy
- jealousy

- worry
- irritation
- frustration
- lie
- resentment
- pessimism, etc.

POSITIVE FREQUENCIES:

- gratitude
- appreciation
- love
- joy
- peace
- tranquility
- faith
- confidence
- transparency
- optimism
- satisfaction, etc.

How the thoughts generate sentiments, you are important to learn how to identify as we felt ourselves since of this the reality that we are creating depends .

Affirmations

Affirmations are not more than positive thoughts.

They beware in our conscience giving the idea of a purpose to the mind.

As some thoughts are irresponsible, they can exercise the control on us of a deliberate form many times going against our ideals; He is like if we have an audio intern repeating : «I am not too good stops ...», «I am overanxious », the money does not conquer me, etc. It is that's why to affirm something the best form is of controlling thoughts in order that they act on our side.

Since we were born we are imposed to listen to our parents's affirmations, our friends, our society, that they get engraved in our irresponsible person and than added up they form the result so that it is our life attracting what in reality IS NOT GOOD FOR us at present.

Many times you use to turn out to be us somewhat complicated to have thoughts positive, unknowingly they help us to Affirmations to change the vision that we have of ourselves, of our environment, of any situation in particular, etc to goad what really we wait for of her to our life.

The principal idea to do these declarations, you are to pass the positive side all the negative side that we have filed in our conscience, that way for example, if often you think that you are not good for any activity, you begin to elaborate positive judgments related toward that that you consider a weakness. You can begin by doing one

List of what you do not want, you do not like that, your Weaknesses since that is something that you know very well and that next begins to elaborate another list that the counterpart would be of whatever you do not want and that you make it impossible to achieve your objectives.
Use the affirmations like if that that you want to manifest, be going on right now in your life, I do not eat if it be something that comes in road or like something that you be going to come in a future. Use the present tense to elaborate your personal list and not attempt extending throughout many things in the same affirmation, since these must be short, convincing and being somewhat focused always very specific in order that they be effective .
From now on, you will be able to find a list of affirmations of different themes in order that you make a choice them what else like you (the declarations that better are adapted what you want to goad), next make your own list. You look for a tranquil place where there not be a lot of noise and read it 2 up-to-date times, concentrate on each of these declarations. Remember what's most important in this process is to have FAITH and when one has ready FAITH reading not like a set will be of petitions but like a series of achievements or full objectives. You are for this motive that we deem the sensation of gratitude as.

Here 6 classrooms of fundamental affirmations

Affirmations for Love:

You choose the positive sentences what else like you!

Make out your personal list of affirmations for love and read it two up-to-date times.
(once when you wake up and another one before going to sleep)

Ya sea que tengas tu pareja o que quieras tener una nueva persona en tu vida, estos decretos positivos son los adecuados para que tu relación sea cada vez mejor o para atraer a la persona que consideras indicada para ti. Piensa, siente, visualiza a esa persona que estas imaginando mientras lees tu lista. Aquí contamos con una herramienta muy especial: LA IMAGINACION. Debido a que la imaginación es creadora por excelencia de los SENTIMIENTOS, tenemos en nuestras manos un poder incluso mas grande que nuestro poder de la voluntad. Siente la alegría de estar AGRADECIDO por esa relación que tienes en mente, siente la tranquilidad de saber que esta lista de decretos te dará lo que estas buscando. Recuerda que podemos estar agradecidos easily cuando leemos con FE, ya que la fe es la certidumbre de que VERDADERAMENTE tendremos lo que todavía no se ha manifestado. Debes ser constante al hacer este ejercicio, tratar de leer tu lista todos los días.

- I am a sincere person to love-
- I am always grateful to love and to be dear
- I have my ideal couple- united, we deserve to be happy
- the confidence underlies our relation
- we shared the adventure of the union of our lives- our love flows like the water
- we created a harmonious cohabitation side by side- we accepted such and as we are
- we are happy with our relation
- we have enough time for us

- we are full of joy and we expressed it every day
- we shared our dreams- we have the energy and vitality share our lives.
- I am a sincere person to love-
- I am always grateful to love and to be dear
- I have my ideal couple- united, we deserve to be happy
- the confidence underlies our relation
- we shared the adventure of the union of our lives
- we have a relation of durable love
- we merit the best of our relation
- we are made to one another
- we enjoyed our company
- our life in couple is marvellous really
- we expressed our sentiments without difficulty
- we maintain the running passion's flame
- we were always full of inner peace
- we respected ourselves mutually
- we have a tranquil attitude and patient in the deal
- the love comes to us easily and effortlessly
- we enjoyed our relation- our love is unconditiona
- we are willing to advance united
- we are happy sharing the time side by side
- everything is all right between you and me
- our lives are full of love
- we felt complete and complete with our union
- we shared our love with our loved ones
- we counted on support and collaboration of ourselves
- we are individual beings sharing our lives
- we reasoned with love
- we believe faithfully in the love like base of our relation
- the happiness invades each minute to have one another
- we thank God to be together

- we have the kind of relation that we want exactly
- we obeyed our personal subjects
- we want the best for each one of us
- we felt realized like people
- we radiated love and liking where we want us to go

Affirmations to attract Money

You choose the positive sentences what else like you!

Make out your personal list of affirmations for the riches and read it two up-to-date times. (once when you wake up and another one before sleeping)

- the money comes to me easily
- I am a sincere person to receive
- I receive the money that I attract with joy
- the riches come to me in abundance
- there is quite enough money for me and for the other ones
- I have the money enough to live in abundance
- all that I see to me around is riches
- I deserve to live well economically
- I enjoy the privileges that my riches provide me
- printed has money at this moment for me and for the other ones
- it is easy to attract abundance
- I fulfill my goals and my objectives

- new opportunities are to the reach of my hands
- I share my money with those who they need it
- I can travel around the world with my money
- the riches are my ally and my lady friend
- I am like a magnet for the cost
- reducing abundance
- I have unlimited abundance
- the universe is unlimited and has riches for all
- I live in an unlimited universe
- I attract the money toward me
- I work with passion and I believe in my goals
- I choose to live an abundant life
- I love the work that I accomplish that abundance
provides me
- work little and receipt very much
- to be rich it is good
- I feel grateful with the unlimited universe
- I use my riches wisely
- I receive money of unexpected sources
- the money circulates freely never
- little I demand being take rich
- the money comes to me like a blessing for me and for
my family
- the universe rewards my correctly made work with
riches
- I invest with wisdom
- I have a good time expressing riches
- I love the money
- my present is perfect
- I celebrate my achievements
- the new riches sources arrive toward me
- I am riches
- I am always grateful with my finance

- I am financially independent and solvent
- THE money flows toward me like the water
- I think big
- I was born to be happy, free and enjoying my riches
- I celebrate the cost-reducing triumph of them besides people
- I focus on the opportunities
- I admire the successful people
- I attract defensive investments
- the money that receipt is bigger I spend of
- I attract the best opportunities to have riches to my life
- today it is my better day financial
I enjoy the luxury that my riches provide me
- I open my mind to receive benefits
- I was born to have riches in abundance
- the riches are a form of energy and I receive it and I accept with gratitude
- I receive money of different sources
- I have a good time saving and investing
- having riches allows having longer to have a good time with my family
- I learn from all my financial experiences- the money runs for me
- I love my work
- if that way I create it, he will that way be
- I invert part of the my profits

3) Affirmations for the on-the-job success

- I am a successful person
- I am open to receive all of the success that the universe has for me
- I offer a product valuable service
- I am happy doing my work
- my abilities in my work improve every day
- I learn every day better strategies
- I am an accomplished person
- I know very well my objectives
- I benefit every day more economically with my work
- I share my knowledge with the other ones
- my riches are my work's fruit
- I do what I like
- success is a state of nature never
- I have the capacity to succeed with my projects
- I deserve to succeed and to live a successful life
- I share with other ones my knowledge
- I know very well where I want to arrive
- I believe in the abilities to accomplish my work
- I feel proud of the success in my work
- I have the perfect work for me
- I attract new opportunities in my work
- the adequate people work with me
- I am admired for the people to me around
- I admire the successful people to me around
- I am always grateful from being a successful person
- I enjoy my work as it were a game
- where always the doors turn on step themselves
- I am receptive to the learning to do my work increasingly better

- I am the successful being that I want to be
- I love my work because you are highly productive
- I have the capacity to do all that I set myself
- I always finish my projects- I was born with success's seal
- I am efficient in all my works
- I am what I consider, as soon as I choose to think about success
- I pay attention to progress
- the success stems from my knowledge
- my life is successful in all aspect
- the success comes to my life in infinite forms
- I am creating riches with the success of my works
- I have the energy enough to accomplish my work
- God is my member and my guide
- I always find the person indicated for each situation
- my ideas are bright
- I enjoy my triumphs contentedly
- I am the lively example of the success of triumph
- I go on the increase with my projects
- I am creative
- my success is a blessing for me and for my family
- I am moving along with wisdom the best road
- my thoughts are attracting more success for my life

Affirmations for Abundance and Prosperity

- I attract abundance to my life
- infinite goods flow toward me
- I choose to open with prosperity
- I am a canal of prosperity
- I have a good time expressing prosperity
- prosperity is my state of nature
- the universe is infinite
- I am always grateful with my prosperous life
- I am opened in abundance to new sources
- I am Abundance, I am prosperity, I am joy
- I elect the abundance never
- I merit the abundance
- my life is abundant, prosperous and perfect
- there is a goods universe for me and for the other ones
- I am happy to be plentiful
- I share the goods of abundance and the prosperity with other ones
- all that I want gets to me limitless
- I am open and receptive to new sources of abundance
- I am a magnet for abundance
- I have luxuries to me around
- I am a privileged being
- my thoughts and actions conduct me in perfect address toward my desires
- I am thankful for the unlimited source of abundance and prosperity
- I accept the prosperity freely and openly

- I am always in the correct place and in the nick of time
- I live to image and similarity of God's divine abundance
- I have all that I desire
- I allow to myself prosper myself
- I have more than what I am in need of in each area my life
- I am solvent
- I think big
 - I am a magnet for abundance
- I enjoy the privileges that the abundance generates me
- I receive all of the goods and the abundance of the universe with joy
- I love the life and I accept my abundance unconditionally
- I am seeing my dream come trues
- I express my prosperity freely
- I share my prosperity with the world freely
- I elect the determination and the persistence to create abundance
- I deserve to have financial abundance never
- my mind is a powerful magnet of productive ideas
- I am attracting big sums of money
- I create abundance easily and effortlessly
- infinite goods flow toward me
- I allow to have abundance and to have a good time of her
- my prosperity is unending, inexhaustible and immediate
- the forms that I attract the abundance in surprise me
- I radiate infinite abundance
- I walk toward the prosperity with love and wisdom

5) Affirmations for Health

- I live a long life and healthy
- I eat healthy and nutritious
- I have a strong body
- I radiate energy
- I am healthy
- the health is shown at my body's each cell
- my physical and mental energy abounds
- I feel good
- I feel full of energy and fortress
- I exercise daily
- I look nice
- I enjoy the health food
- I have my ideal weight
- my life this fills joy with joy
- I feed my body, my mind and my soul well
- I am nice to my body and my body is nice to me
- daily I go away feeling but healthy
- I exercise and I obtain results
- I keep watch over my body with love
- I am always grateful with the health of my body
- I have good habits that provide me health
- I like to face me the mirror and to see a healthy person's reflection
- I breathe fresh air that makes me feel well
- I live a quiet life
- I have a good night
- I have fun healthily
- everything is all right in my healthy body
- my body is the divine expression of the perfect health
- I feel gratitude for my radiant health
- my body is source of health, integrity and force
- I become every day that happens but healthy

- all my body vibrates of health and fortress
- I have a good night at night and I enjoy my energy in the daytime
- my vital energy leads me to toward success
- I eat in adequate proportions
- I live a tranquil and relaxed life
- I am agile in my movements
- I have fun healthily
- I meet with positive people
- I embrace positive attitudes in all aspect
- I respect my body
- I have a good time when I go out to walk
- I feel happy
- I recognize my fortress my healthy body's product
- my health stems from the harmony between my body and my mind
- my excellent health is a motive to celebrate
- I radiate strength
- I feel good, I look nice
- I am vitality
- I am an abundant being in health
- everything works well at my organism

6) Affirmations to be happy

To find the happiness is task of all. We can be happy at this moment if that way we decided it. We quoted a very valuable part of the book Conversations with God where the paradigm talks about reverting HAVING – DOING – somebody BE. ¿That does this signify? Let's observe :HAVING : We believe than if we have longerDOING : We can go on vacationBEING : And being but happyAnother example would be :HAVING : We believe than if we have money DOING : We can do good businessBEING : And being richerHere it is where we are failing because we can be right now what we want to be. Simply elect it. Say :I CHOOSE TO BE HAPPY (Also we can choose to be plentiful, to be patients, being successful etc., But we are going to focus on happiness)The first step is to revert the paradigm as it should be: BEING – DOING – TO HAVE.When taking your decision, you are going to feel good, feeling good you are going to be a person but kind, but loving, but focused in the good things that you can bring a day. You are going to act better: You are going to MAKE what you sense that he is adapted for your status of HAPPINESS: You are going to treat people better, you are going to have a good time doing your work, you are going to get time to have a good time with your family, your children, your parents Cuando your attitude changes, the results change also You Are Going To receive your part of your election to be happy because you are going to pick up your harvest's fruit, hour becomes of HAVING . To have is To Receive: Receiving the result of the way you felt yourself in and you acted in because you will see as you will bring you, the good things that you have wanted to HAVE.

HAVING does not produce TO BE, but BEING if you produce TO HAVE. To affirm is to elect (deciding something). You can find a list of affirmations to be happy here. Organize your personal list with the affirmations what else you are interested and read it every day. Remember than:

The note is the virtue for her than everything them besides virtues yield their consequence. Balthazar Gracián and Morales.

- I am happy
- I am joy
- I am God's perfect expression
- I am all right and I improve every day
- I feel happiness
- I recognize the wonderful being that I am
- I enjoy the present measureless
- I am happy to be who am
- I am kind to the other ones
- I thank the good things that I have
- I advance with certainty because I know that everything is all right
- I accept with love my difficulties and I change them with wisdom
- life is a party
- I wake up thanked
- I appreciate the people that live with me
- everything is all right never
- I am a successful person
- I live the present
- this is one of the happiest days of my life
- I love to be the person that I am
- I am always happy from being healthy
- I am full of energy
- I radiate happiness for where I go by

- I have thoughts of happiness
- I am always grateful
- I am a beloved being
- I share with the other ones
- I appraise my qualifications
- work with love
- I respect the other ones
- I enjoy the success of the other ones
- I choose to be happy

Gratitude:

Gratitude is a sentiment. We can feel ourselves thanked having positive pansies toward what we have, we do that and we are that. This sentiment puts us in a better position to ask it to the universe whatever we want to manifest in our life.

You are not simply to say thanks, but having that sensation of happiness to give him a value to our life, to be convinced that what we have is something that makes us feel well because to feel to us of that form we see like the gratitude converts itself on the way over to attract over that that right now we have .

Many times we focus our thoughts in that that We Do Not Have, so that we need and we felt uncomfortable that's why, that's where we sent a sign to the universe that attracts over that that makes us feel badly. How many times as we woke up thinking about the debts, in the fears that we have, in our

Incapacity to get better opportunities? We focused on only that little bit of negative things that has our life, but we did not look further on where all positive that we have meets : Do you have children? Do you have parents? Do you have a place where living ? Do you have a job? The key this in marveling at what's quotidian, in that that the unnoticed elk passes your eyes, but that you have the reach there of your hands.

You thank the awakening and if in the morning you stick at that positive sentiment, keep on thanking, and at night ... you thank a little more. You thank God for the universe, for the flowers, for the air that you breathe of unconscious form. Thank for the simple things and you thank also your big achievements, thank for your incalculable capability to obtain that that you desire and that establish that happiness that it invades you to be grateful, you feel that comfort, that peace, that freedom ... all those positive sentiments that the gratitude generates you.

Make out a list of the things you feel gratitude for, you can begin for looking about oneself to identify the people that you want that are to your side, identify also your achievements, your faculties, your health ... You Include also what's material, your belongings, that that you bought, that they gave you as a present, that you inherited . Use your creativity and what the land provides you thanks also: Mountains, water, the stars, flowers, your pet! Read your list two up-to-date times and you enjoy the sentiment than these generating, that reality than these creating .

The one that the Mr. to marvel neither of getting enthused more does not possess would be worth being dead, because its eyes are off . Albert Einstein (1879-1955)

The wonder of a very snowflake surpasses one million meteorologists' wisdom. Sir Francis Bacon (1561-1626)

If so, keep on reading this article where I expose your three secret steps to be applicable, successfully and rapidity, the law of attraction in your life.

secret # 1:

Seat what you desire and a great deal!
It is not only issue of in that we considered to get what we want, but also that we felt with regard to this matter. In fact, you would go so far as if to say that our emotions are a lot more fortress and powerful than our thoughts! Why? Good, we can think and thinking about what we want that it happen in our life, but if our emotions are of fear or we felt too intimidated by our desires, then everything is in vain .
The universe has the capacity to read your emotions.
If you want something to happen, you must make sure that your heart is on line with his goal. For example, if you want to earn the football game tomorrow, he needs to feel motivated, enthusiastic and to feel really like a winner.

secret # 2:

All are vibrations

Somebody has told you is the fact that you are a vibration? One of the beginnings of the law of attraction that not a great many people render is attention to.

Basically, everything that they exist they are not more than vibrations, I solidify the material you do not exist as such, only it is energy, and energy is vibration, your body and whatever you surround us, this constituted of atoms that they vibrate .

Our thoughts and sentiments are vibrations. These vibrations are sent off to the universe that is also a great vibration.

All that thinks or feel you vibrate and it resounds with the universe. That way he is how the universe is paid of our desires, in reality we are a part of the whole and it is that's why that you have to be aware of his thoughts and emotions because that whole will resound with you.

secrete # 3:
Writing affirmations gives you an extra impulse
A little-known law of the secret of attraction is to write affirmations. ¿Why? ¡Because be ten times as much powerful! When writing the things, a conscious effort to think about something is done .
Why do you believe that professors always have requested that they take notes during the classroom? He is not only why next have him something that to make a revision when it take place home.
He is owed to write the things reinforce the lesson in your mind. Writing this they do that they record themselves more deeply in your subconscious and this have more potency to the hour to work emitting vibrations to the universe.
And finally, an extra child, for example you have learned three concepts what if you learn in this article you will be able to increase the power of your efforts with the law of attraction and improving your life, now you have a knowledge what if the coparties with the other ones will also enrich their life and, you will curiously see that your life will be seen enriched more and more.It hits and you turn your edges like everything multiplied!

Be you gotten interested in the use of the law of attraction? Because good In Order That you be efficacious in his life, there is to have very in account these six essential elements that you must apply in his life.

1) Patience. The law of attraction is very efficacious and powerful, but not always it works of instantaneous way. If you wish to obtain real and powerful results, it must be diligent and patient to work toward his goal at the same time that you attract toward you with his positive thought. In a way, you and his objective has to meet downtown. In short, you have to demonstrate a positive attitude and patient toward your desires, without demanding than that million dollars or that dreamt couple, appear in your door in the next five minutes, be patient and persist .

2) The correct desires: This you sure that do you know I that you desire ? And that those are the desires that really you desire by heart? A great deal people apply the secret without having what they desire very clear, spend time in thinking up what you desire EXACTLY, he will feel that but caning to attract him. Besides clarify if what you want to attract is what you desire really, really does it want to be rich? Yes? Would work 14 up-to-date hours to do it.¿Probably not! Then perhaps do you want to have a pleasant and abundant life, not so? If his objectives do not make sense with his attitude, so that you dictate the heart to him, hardly you will get what you want, you petting sabotage him before achieving it.

3) An open mind. You also must have an open mind. The law of attraction says that a superior force in the universe is going to be who send his desires its way. You may have drawn a plan of like to achieve it, but things do not happen like you you arrange them, however an opportunity can come to it and join unexpected place or he can receive something else that you were not asking for but if you desired and you needed . Do not be strict and open your mind with a world of possibilities and you will be surprised so that it happens in his life.

4) Look for spiritual objectives. You can attract with the law of attraction everything s the material desires that you want, and when you obtain them, will put an end to the day a sensation of empty hardly explicable however the spiritual goals, the desires than go they will make more sui's wing same for him to have a complete and happy life: As soon as you did not sole ask for for you, else you attract and work for the other ones, for an objective that you benefit a great many people or the enclosure look for relatives's happiness and friend. Then his happiness will be complete.

5) Look for in your inside. The law of attraction requires a journey of discovery and learning. You have to learn and to experiment to can of the law in his own life. The common beginnings can be the same, but the techniques the fact that they worked for other ones you are able to be out of service right now that people have very singular desires and peculiar characteristics.Many try to use the secret to try to subdue the world and that you become adapted to his desires, that simply out of order, first have to dive in our inside and conscious being of our limitations and working

with them, discovering our weaknesses and our fortresses and making good use of the tools that are us for innate, to find the pair will be ideas impossible if we have an impossible personality to try, to construct a good reputation and a good cost-reducing position will not be possible if we are indolent and little determined if you want to change the world, change first on the inside, take an attitude but positive print and determined of life and you will see like the world change your surroundings, he will not be like your you demand to that he be, it will be in a way that it will result still better from what you create than of letter b being .
You apply these small pieces of advice when you apply the law of attraction and you enjoy a life of dream!

More methods about law of attraction

1. Ask for what you desire:

The first step to utilize the law of attraction is to ask for what you want . I can hear somebody telling you: "That is easy, I am going to go down to order all that I want !" But really it is so easy? You are safe that those desires are the ones that you really want or are they desires that have been dictated by the other ones? Sometimes both concepts blur out . It is important to have what you want obviously. It is so easy to think that we want something when in reality Creemos is a duty that something must be wanted. We believe that the other ones want us to do something. Once once this was clarified, take your time to write what you want with the bigger possible detail. When one has clarity about what's great that is what is desired, easy to use the law of the attraction to get them and to can to center more in his objectives.

2. Living the Life with Purpose:

Many times we became careless of the for what of our acts and our desires, what you maintain us motivated is for our intention being enough. The scribe why wants what you want . What does it obtain? ¿What sentiments do they have him when you come to his intention? This why will maintain you on the move forward and you will help him to filter his activities. You will meet asking around: "This activity will bring closer to my objectives or on the contrary do you drive me away of the? Put one great " cause " of an intention and that you will goad rapidly to his life.

3. Step by step for your desire:

The following step is to feel the sensations that you will have when you come to its desire. The law of attraction responds to our sentiments. Visualization is a wonderful form to begin to feel those sensations NOW. Take each by your intentions and imagine that you right now have them. What you feel when coming to his intention feels . Put the color, the sound and the smell to the image. Ahh. just enjoy it!

4. To clean the mental disorder:

We all have beliefs that they lived either conscious or unconsciously. When we mean to earn more money or financially free being, for example, we found our beliefs on money hastily. "¿Who does it believe that you are for winning so much?" "Money is difficult to find." These are the beliefs that will be put at your road to attain his intention to be financially free.
But guess what? ¡You can elect your beliefs! ¿What beliefs do you need to arrive to his objective to be financially free? " The money enters never easy and effortlessly." Or " The financially free being is easy to achieve." Elect the beliefs that operate for you against you!

5. Getting going:

I am in accord in than the knowledge of how utilizing the law of attraction does the easiest life, but that's not to say that we can take a seat in the sofa and we waited than our desires fall solos in our lap. ¡Still we have to take action! When you are visualizing what you want listens to the ideas of action that they may show up. Think as you would be able to achieve what you desire and put hands to the work with it, if you concentrate your intention in achieving it you will see that he puts on his universe on your side in order that it attain his fast goals.

6. Fall in love with your present-day reality:

The most powerful step that can take to start with to attract what you want is to be grateful.Look about yourself and feel grateful by the good things that right now you have in your life. Be thanked for the beauty in the nature that you have, for the good good eaten food, the accountsIt will help you to this to get a better life! Every time that you focus on what you like of your life you will bring over those wonderful things toward you to you.

7. Celebrating!

This is my little favor! To celebrate his successes is a great way to focus on what you want . Remember in celebrating the small successes, as well as the big.People's majority is not really conscious of the achievements that it accomplishes to I deliver it of the day: We achieved what at work, the tasks at the home and with the family that we have that we can not give, the good joint moments, the small things and those whims although he be once in a whileYou celebrate the little things that right now are in your life and you will see like each time you have causes for celebration more and more.!

More about law of attraction

All that vibrates in a similar way is attracted :

In this material world, we have the gravity of the land that attracts all of the objects toward his center. If I throw an orange up, I can be sure that soon you will fall .

In the dimension of the vibration, something similar similar succeed . I give a vibration and soon it return myself, but in this case, it return myself toward me.

I have listened the aforementioned popular than rumor has it that violence generates violence and that silver pulls silver. Everything goes back . If I give hatred, I will receive hatred. If I give love, I will receive love. If I vibrate with hope, they will open doors to me. If I vibrate pessimistically, they will close doors to me. I will manage to understand with this Law of the Attraction for what this is everything he is that way.

The effects of this law are fascinating . I can observe how the intellectuals assemble, the rockeros frequent the same places, businessmen themselves the commerce cameras or industry and union members instruct their labor unions. The lovers of the popular dance tend to form a group, the lovers of nature feel german and the same thing happens with the drug addicts, the ones that like the opera, the fanatics of soccer, the hunters, the children and them fevers of the Internet.

All that vibrates in a similar way is attracted .

I attract toward me all that vibrates in a similar way my vibration:

If I vibrate with fear, I attract more factors that they will make me feel fear. If I vibrate with sadness, I attract more reasons to be sad.

If I vibrate with joy, I attract more circumstances that will make me feel and expanding my joy.

I will attract more thoughts that they will cause joy to me, I will attract people that are happy and will increase side by side the joy of each one. They will begin to happen to me things that they will feel me more and more joy.

A friend that was very indebted asked me:

" That I understand you and I like it, but how can you explain that I should have so much debt, if what I want is all the opposite? "

You say that with your words you want the opposite, but your vibration is of scarcity. You vibrate with without money, you get upset when paying your accounts, you get angry because you do not have enough money. The sign that you emit constantly is of scarcity and that is what you receive .

" But that is not what I want !"

It is certain that you do not want it, but the sign that emit and is what is returned to you is that one.

We will understand how changing that later on. For the moment, acquire conscience of how you are vibrating on the inside, before the words – with your thoughts and emotions – and you will see that that way he is .

What's certain of the point is that I attract toward me all that vibrates in a similar way my vibration.

My vibration is the mixture of my thoughts with the emotion that these generate me

Vibration popularized the term itself in the sixty's decade. It was said than if a person had a form to be nice, you vibrated very well. The people wrote themselves like of good or you vibrate of mailbag.

You vibrate her of a person his form comes from being and thoughts and emotions are the raw material of the vibration.

An optimistic, nice and constructive- attitude person, vibrate of positive way. His thoughts and emotions are harmonious, his presence is radiant and his company is beneficial.

A person of mailbag vibrates you have thoughts and negative emotions, his attitude is pessimistic and his company will be undesirable for destructive.

But beyond the generality of being, I can vibrate well with mathematics and evil with the letters. I can vibrate well with work but fatal with love.

For example, if I think that I am good for mathematics and I sense that I move without difficulty through numbers, then I vibrate in accord with mathematics.

If I think that I am not very attractive for the women, that they always are going to prefer another men and that I am not going to manage to establish good relations, I will feel a very ugly emotion.

My vibration will be of handicap, of incapacity in front of the women.

When vibrating that way, I will attract over the same thing, because all similar is attracted .

I will attract more thoughts of incapacity, I will attract the rejection theirs, I will attract friends whom you happen to them it per se and at the end, I will construct a future with those characteristics.

Everything this happens for the way I am vibrating in and that my vibration is the mixture of my thoughts and my emotions in.

I am a constant emitter:

I emit a sign constantly. Each thought with his correlated emotion is a sign that I emit .

It's possible that you not have a lot of conscience of this, but daylong I pass you vibrating one way or another. That happens normally and you succeed everybody.

For example when I get angry . Emotion is horrible and dried several minutes and sometimes hours and to complete days in that I do not bear myself neither myself.

I do not have conscience that to vibrate that ugly I am sending an every bit as ugly sign and I am attracting more and more therefrom per se that you bother me so much.

On the contrary, if I am happy and carefree, my vibration is delicious and the sign that I emanate is equal of good. Although you not find out, I am attracting reasons to be content more and more.

Although you not find out, I emit a sign constantly.

I am like a radio station that emits a sign and that sign is perceived.

God, the Universe, the whole you perceive that sign and immediately you respond

This is the same thing as The Biblical appointment ask for and it will be given to you. The thing is that I always believed that to ask for era with my words but I already understood that to ask for you are with the vibration.

When receiving my sign or vibrational order, the law of attraction immediately begins to join my sign with all that is similar.

The Universe is full of capable resources to make draw with that sign that I emit and fulfilling my order.

If I vibrate with the desire of abundance, that very thing will give me the Universe. But if I vibrate with fear, equal, I will receive that very thing.

The Universe does not wonder if you are something that I want or that I do not want . The Universe, obey my free will, my election, and respond immediately.

The fact that I be sincere to receive this is another matter . It may be that I ask for something but do not make it to me or that you have fears be more than enough that is going to generate never.

If for the reasons that they go I am not open to receive what I have asked for, you will not become manifest never. This point we will approach him extensively in the Permitir's Law.

For the moment, what you interest to know is that I emit signs constantly and that all those signs are perceived by the Universe and responded immediately.

My vibration attracts similar thoughts ...

... people that think in a similar way and events that are of the same sort.

Normally the attraction becomes manifest first with it subtler and matches getting to the point step by step even concrete.

Then, thoughts of the same frequency of my vibration are the first thing that I attract . If I keep on vibrating with a specific.

If I vibrate with the intention of earning more money, the thoughts that I attract can be very varied but of a clear address.

For example, " I appreciate the abundance you live in such person; The freedom likes doing me and to have what I want ; That company will open soon an office at her city and I would be able to win better if I work there; I want to learn how to vibrate in accord with abundance ".

Little by little, I will attract thoughts of that same address more and more and soon I will look surrounded of people that they think in a similar way. On the way in in touch with these people, we will mutually encourage ourselves in the desired address. Each person will contribute something and I will do it also. All these people to me around they will be consequence of my attraction.

To the time, I will have opened many doors in the direction of earning more money.

I will have developed a best-suited intention to achieve it, I will have gotten a series from contacts keys, I will have touched the correct doors and I will be in bigger capability to make good use of the opportunities that turn up to me.

To the time, he will begin to become manifest never a bigger flow of money, be because I achieved an increase in my salary, because I initiated a new lucrative activity, because they will pay me forgotten debts or thousand one forms than neither I can imagine at this moment.

The point is what if I ask for the die will be me for and my order is to crosswise of how I vibrate . My vibration attracts toward me
Similar thoughts, people that think in a similar way and events that they take me in that same address.

In order to comprise the point better, I observe how my life is present-day.

The circumstances of my life result from my attraction:
My relations, my achievements, the things that I have, my status of health, my work and besides circumstances result from my attraction – volunteer or involuntary, conscious or unconscious.
I am what, I do and I have it is because I have attracted you with my vibration.
If I want to know how I have been vibrating, the one thing that I have to do is to observe all that surrounds me, to observe the people that are present at my life, the things that I do and all that I have .
Everything that is that way because I have goaded you to vibrate in a way.
There will be things that I like and others than no, but everything have been attracted by I. I am the only responsible person of my attraction.

In other words, whatever I dislike my life result is attraction to someits:

What makes me angry
What frightens me
What saddens me
What depresses me

I do not manage to accomplish what
I want what but I do not have
I have what but I do not want
I do what and you want not to do
I do not do what but you would want to do

In short, all that is not desirable to me in my life is here because I have attracted you.

We right now saw the example of that friend that was very indebted and was not understanding why you lived in scarcity if he wanted abundance.

In the same way, if I have low self-esteem, I do not have many friends, step mortified by the negative emotions and nobody wants – although it be difficult to believe it – this he is that way because I have goaded it to my life.

You may have begun from my infancy and one may have confirmed in adolescence; He can be than neither remember how happened everything this, but is what he is because I attracted you.

What's important now is to understand how it is that I attract toward me, so much that is pleasing to what I dislike like. Later on I will explore the way to give the subject turn and to begin to attract only what definitely I want .

The circumstances of my life result from my attraction. Since I attract in show of how I vibrate, the conclusion is that my life is what he is for the way I have been vibrating in.

With this you finish all of your story that of that there is a destination right now drawn for me. My destination I shape it myself and you are my thoughts and emotions, through my vibration that I do it.

The vibration more than the words they are my source of attraction:

My words not always express the way I am vibrating in with precision.

If I say : "I want to be healthy " but on the inside I am with a great fear to fall victim to cancer, the truth is that I am vibrating with cancer.

If insincerely I say : "They are going to burst with in the exam mathematics ", but in the background, I know that I manage the matter well, it's probable that I be doing all right .

What I say not always is a reflection of how I vibrate .

This is evident in relations. If I tell my couple that I want her because my part establish it, but in addition they cross me a heap of thoughts on his way of being that I dislike, she is going to feel and energy is not going to flow.

Children are specialist in reading the vibration.

If for the reasons that they go I do not grant permission to the girl to do something but on the inside I sense that it would be fantastic that I may do it, she will perceive it. You will not understand the no and you will keep on insisting until you achieve it

There are oceans of examples that my words do not reflect what in the background I am vibrating in.

To me what voucher here is what if I ask for will surrender . But it will be given to me relating to my vibration, necessarily do not give agreement to my words.

I right now have of course ...

I attract so much what's positive like the negative side:

What's important now is to know how do I make somebody acknowledge to if I am attracting it plus sign or the negative side?

The answer the emotion gives her to me.

I will know if I am attracting what's positive because I will feel a positive emotion, because I will feel rich when vibrating that way.

When what I am attracting is negative, I will feel a negative emotion, I will feel ugly.

This I will understand you very easily, because you succeed me many times a day.

When I vibrate tastily, putting my attention in the in-fashion beauty, in how much I appreciate certain people, in the happiness to do what I do, I feel rich.

When I complain about my work, I renounce from the preys of transit and I sense that I am surrounded by disagreeable people, I feel ugly.

When I vibrate rich, of positive way and optimist, I am attracting over everything that that makes me feel so rich.

At the most intense be the emotion, attraction will loudly be more.

I can feel a very intense emotion, but he is positive or is he negative?

Attraction is every bit as intense so much if I desire that you are to have something or of not having something.

This is all this theme's most delicate point.

If I feel a strong desire to achieve something, I will attract that with more intensity.

But I annoy what always I have believed, while intensely YOU NOT WANT more something, more I will attract you also.

You seem absurd, but if I explore it from the perspective of the vibration, you are going to prove to be logic.

If I vibrate intensely with the desire to attain the cost-reducing abundance, to have money and to do everything that I can do with that money, I will be goading that very thing to my life.

But if I vibrate intensely with despair that money is not enough, that I do not generate enough, that I can not pay things what else I want, I will be attracting more scarcity.

This can compare it to the quests that I do in Google or Yahoo in the Internet.

If I ask for Google that you give yourself the pages related to I want more money, it will return a great quantity of pages that they treat on having more money to me.

But if I ask him for the pages related with money it is not enough, you will give back a great quantity of very different pages to me. I will find the themes related with scarcity, debts and poverty in these pages.

Still further, if I ask him for the seeker minus debts, the key word keeps on being
Debts and you will give back the pages with that contents to me. Pages will be similar if I type I want debts or if I type I do not want debts. A yes or a no by the front are not going to do bigger difference.

That way he is like works this subject.

God, the Universe or the whole, respect the freedom that I have somebody elect. The theme of free will has been established in several Sacred Books. Swim neither nobody is going to correct my order half-way down the road. I will be given such and as I ask it for.

I am free to ask for what I choose to ask for and the language of that one to ask for it is my vibration, the words than use.

If I vibrate with scarcity, I will be given more scarcity. If I vibrate with abundance that he will be what you receive .

If I vibrate with that nobody wants me, I will attract more lack of affection. If I vibrate with that I want more love, my attraction will be on the right track .

If constantly I feel horrible emotions, my vibration is that one and I will be attracting over the same thing.

I must begin to vibrate with thoughts in order to attract the peace, the joy and the calm that they take me in that address.
The solution to all that subject is here.

The Ley of Attraction does not correct my order. When I am vibrating with scarcity, the Universo does not stop the subject and it says : "This man is asking for scarcity, but I know that what you want is abundance – I go to abundance to have authority." That would be to interfere with my free will.

Of that absolutely I can be sure you come from that the Law of Attraction equals that that is similar to my vibration, without interpreting neither correcting nothing with all.

It is my responsibility to elect how vibrating . I have to vibrate with that in order to attract what definitely I want that so much I want .

If see it from the other side, in order to stop to attract everything that that I do not want, stop vibrating nevertheless that I do not want .

The following course is precisely about this mechanism.

How do I attract my life whatever I want ?

At the same time, I will understand to depth how it is that I stop attracting whatever I do not want .

Well, with this I complete the basic exploration of the Law of Attraction.

The Law of the Creation is the name of the following course. With the tools that I will learn to use, I will be able to straighten up the address that takes my life, I will impulse myself toward more desirable statuses to be and I will know than the life of my dreams, more than a dream, you are something completely reachable.

Law of attraction to be rich:

Money:

An illusion, a shade of a little more
The first step toward having riches is to know that he is .
Few people know what really the riches are as such same.
¿That it the riches? ¿That it what the riches cause ? ¿What
is does that cause the cause of the riches? Let's begin with
money, the riches's worldwide symbol and after let's
deepen more.

Money is not real .

Money is a legal instrument, a form of interchange simply.
We used it to interchange value. Represent value.

Money is the value's body. The physical performance is of
the value that you increase and that descends in ourselves,
within us. No within the things out of us but inside our.
¿That would you come from the value of the things, such
like a car, but out of here for us? Swim, at least for us. In
other words, we, observers are, them that we put value in
the things. The material things you do not have monetary
value in himself same, we give it to him. Therefore, money
is the physical performance of a particular sector of our
internal value, inside our, within you. For that reason ó is a
house a company's actions that today are valued in one
million dollars can incur in, tomorrow, their valuation in the
middle when you introduce the fear in those hearts that are
involucrate.

Fear a portion of the internal value of the participants kills and this is reflected in the paper money that is the value's body.

Here there is something more: The paper money not even represents money completely. Reasonably, you can not represent it. 4 % of the money that is in the benches exists like cash money according to some (that they vary according to the country) very esteems.

Imagine how much cotton fibre, linen, the pulp of paper and metal would need itself to manufacture all of the money that everybody has in his bank account. Imagine as much as it would take to space to store all that money in the shape of paper money. If one have to pile up one million dollars in one-dollar bills, the sink would weigh a ton and it would have a height of one hundred meters. The money not even exists more like gold stock. Exactly for that reason, in the decade of the 70 we ran out of the reasonable capability to maintain a gold standard.

Then, That is he what exists like the money we are talking of?

Good, it is a mass illusion. Music only written numbers paper ó once storage devices of the computers were put in, ó to be but precise, records! In order to paraphrase it, for each 100 dollars ó his equivalent in any other coin, only 4 dollars exist like bills ó coins, in the meantime than the 96 remaining dollars exist like numbers once papers were written about ó data of computers in the benches, companies and another entities.

Last time that the people stopped believing in the large-scale system you were in the Great Depression when great quantity of people slipped out at a run to the benches to withdraw his money. This was not the cause of the Great Depression, but to a large extent, he was what accelerated her.

Therefore, money is not real, a little more is it. Money is only the shade of something else. The first step toward the riches is to know what what really is the money, ó to be more precise, that he is what you represent is . Learn how not to look at the money most of the time. As you will see from now on the day that you watch to the money should be rare of the way in than do it today in day the cash, the bank accounts, costs, etc. This is the shade merely and no what's real. Looking at the shade, the physical money, as you are going to check it right away, is not advisable and healthy no for you and your finance.

Instead you watch the value inside yours, within people and the flow and interchange of this value between the people. Our internal value is what money creates .

The money is the shade to our internal value. Develop that internal value in yourself and in the other ones and your money outside and riches it will increase automatically, without fail.

However have account this: The money represents an aspect of the internal value of a person but it does not mean for me to represent all of the internal value. This is very important . This does not treat of the car assessment itself.

The money represents an aspect of the internal value pertaining to the riches. Therefore you can not say that a rich person has a self-appraisal ó higher value than a poor person, but you can say correctly than in themes that they relate with money, the rich person has an internal value more lifted in that aspect of the value the person chooses to exercise a bigger proportion of her internal value in. Concient of the Riches calls the internal value's this zone that reflects in the external money, when you exercise it herself,. He is available for everybody equally and he can be developed for and within everybody equally. Like anything else that is important for the maintenance of life, like air, the conscience of the riches is free for all. One can choose to unroll her ó not, to drill her ó not.

At any time you can change your election and you can stop you nothing short of you.

You need not nothing of you to increase your conscience of the riches and therefore money. All that you need this within you right now. Maybe you had forgotten but you are there. You are going to remember it and the first step is to remember provided that money is not real ; The shade comes from a little more.

And there is another secret here: The conscience of the Riches is simply to your the expansion conscience and the perception of the delicious parts within your being. For that reason it is that all that you need to increase your Conscience of the Riches right now is in yours. You right now are rich, but they taught you how to elect not to experiment.

your riches. Everything changes this knowledge. Just as the rich do it, you can know now like doing it and beginning to experiment the rich in you.

You have bigger capability of riches inside yours of the one that you may experience in all your time to live. You do not need to get worried if you have caught up with your limit to make a fortune somehow in particular ó due to any circumstance. Not even you have to know like to turn the conscience of the Riches into cash paper money, because, as you will see, it will happen automatically.

All that you need to do is to expand your Conscience of the Riches and to exercise it, to act upon her, being she. To you the situations and opportunities for the equivalent conversion to cash money will turn up automatically. No one of the people that today are rich outrageously they may have forecast, if not they were rich, the exact sequence of the events that led to having a so immense riches. Probably they had a set of goals and a plan but anyone of them can tell you that they interbreeded with countless coincidences and opportunities that they finished to join him the points of manners that they never may have forecast. His goals were his own doing but the roads that they took to his realization and that still exceeded themselves were intelligent incredibly and even so, unpredictable. You will see now like to be done to that that happen to you in your life – perhaps may predict the sequence but certainly you can do than those Fortunate coincidences happen to you every day of your life.

On purpose, not only the paper money is not real . Many of the things that they surround you and that you consider real neither they are real . You are about to embark in a beautiful journey that empowers you and freedom and that you go for showing you exactly that he is the world from a perspective that never before you had had . A journey that you will open you with is eyes and you will untie your wings. You are about to look at that he is what you have below Life's hood; You are about to learn like to adapt your world to your pleasure.

You are about to attain the conscience of the Riches. Once you do it they are going to make you difficult to avoid the success and the riches. The success and the riches will persist to you automatically. Yes, you read well. You do not have to worry about his quest and still that way they are going to find. You are going to be free to experience another aspects of the life that not even you dreamt before, dimensions of the Being of Life and that are really incredible. The same thing happens with happiness and you are going to see also here in this book.

Let's begin now with the concrete theme, if it is that still you continue to be interested.

You are about to initiate a journey which you will know like creating all of the riches and happiness that always you desired, Now, without no limit at the end of. Soon you will see like many eternal truths be more than enough who you are in reality, what you are doing here and of that the game of life is treated. The steps go from the journey that you are taking with this book here:

1. You are going to take a look at the Carnal, because knowing so that you are made and the world it is the first key to know like to find your road. After that, you never are going to face the world in the same way. You are going to have an incredible sentiment to be implicated and to have to can in the universe

2. From now on the secrets of the time beginning with the fact, that the time does not exist are going to make oneself known. You are going to learn like to use this illusion instead of being used by her. Only the Ahora exists

3. You are going to learn like to create your universe as from the quantum field using the imagery of your mind. This is the party of the first part of the lessons of creation.

4. You are going to learn how to create using your thoughts. You are going to learn correct the way to think and in order that she is, in reality, the mind and when to put it out for your own benefit

5. After you are going to examine the true power to put goals of to thousand of a way never you heard in before, of a powerful way

6. From now on the most powerful comes from all of the tools of creation, the state of Being

7. Last tool of creation, the action, it will be for revealed to you at his correct position and purpose

8. You will learn after the magician and certainty's vital ingredient and you will learn like having it in abundance

9. At this point, the moment will become of examining the first Law of the Universe like using her to have happiness and riches in abundance.

10. Still being within that law, we will examine that they are circumstances. This is going to give a crash, is going to make laugh, is going to empower and is going to liberate

11. While we are in the theme of the circumstances, you are going to see as you are, you will attain permanent success and you never will fail

12. From now on you will examine the principal murderer of the riches and the happiness like avoiding it totally and

13. From now on you will move over to step by step bigger things.
You will begin with your purpose for own election here below.
In order that did you come here? You will see it

14. Next the gift to hit and the gifts that this brings in return to you will be revealed

15. From now on the power of gratitude will come to light . This is going to try to be extremely powerful

16. Finally, time will become of examining the conscience, that he is what you make to wake up to you Here, Now

17. And next you become really interesting when you examine your own Being, the first Cause of everything that you have in your world. Get acquainted with your Being and your world will change drastically.

18. After that, you will see that what you and everything else come from is Being, that what's more big integrant part. It is the One. You will with all put knowledge like relating your He Is, the Source and experiencing it to you at a position of immense joy and abundance

19. You will see in the same subject matter how abundant really you are

20. To depart of there, you will examine your real nature and I eat to claim it. The nature is of the pure joy

21. From now on, and closing the circle, you will see like to drive the paper money better, just as you know it, in order to enlarge your riches

22. And to conclude this part of the journey, you will be guided toward the steps that authorize you and than maybe you wish to drink after you finish this book.

How read and understanding this book

The way to understand this book is to read it once completely. As you read you will have many questions and something are not going to have no sense. No matter; Simply keep on reading . The following chapters are going to clarify something than such time ó has not understood before that you thought that they were truth in previous chapters. The idiom is a linear thing, but the conscience of the riches is a totality, a nonlinear totality and therefore I join the step it can be connected with the step seven and that way like that. The conscience of the riches is a status of being while the language is a symbol; The states have to experience themselves and symbols can not represent the experience with precision, only they can show the road, being a guide. Logically, as you read, you will find many incredible things that make sense at that point, but they will have his total sense when you finish the book and have the totality of the in you. He is there when you begin to have you wither them!

Once you read it completely, read it of new DSPACE. You will be able to understand the themes in more complete form since you will have once the totality of the book was incorporated in second reading.

The material that you will find in this book contains many capes of understanding. What you understand today will develop one of your deeper capes of an occult truth of an application of an identification, and when you read it tomorrow. You read a little part of this

I free every day, still after of than finish it the first time and your internalization and manifestation of the riches and the happiness will accelerate .

Something else that you should do is not only to read with the logic mind but also you have to feel the lessons of this book. Something are logic while another one relate with a world where logic can fail to understand but your Ser knows them and you understand them to perfection. As you read this book, keep open, feel the essence of what you read . Many things will be no use for your brain that is finite and of four dimensions. Something are infinite and multidimensional and only your Ser knows them because it is Infinite and multidimensional by nature. Sometimes you will sense that you understand something to a deeper level but your mind can not imagine it and still that way you understand it deeply somewhere within you. Obey the deepest level since the mind, anyway, never will be able to understand it. In any event, the mind is a tool, but all that you know is in your Being. You are a Being with a mind like powerful tool while you not take control on you. However and unfortunately, they identify the people's majority with the mind, they believe that they are the mind and that's where they begin problems and limitations. Now you are going to go above.

You read this book and take it to the heart, live for him and the fact that the riches, the prosperity and abundance arrive to you in shape and I eat quantity that never before you thought up will make possible through laws that never they fail . Take it with you where it be possible for you. Leave him next to your bed and read it during some minutes before going to you to fall asleep. When you finish, start again, some few pages per day. The repetition causes the internalization and it becomes part of your nature.

You will understand this book insofar as you want .

Are you ready? Still do you be interested? Let's begin

Carnal Cuántica: The first key to find your road is to know of that you are made and the world.

Right now it must be asked that you have to see the Carnal Cuántica with the riches and the happiness. Well then, you are better than you create it because you have everything to be related! ¿How can you build a house if not you know of that it is done and how does it be built?

The Quantum Physics begins to explain how everything in your world happens .

You are capable of manipulating all your physical world directly but it may be that you not become aware of it. Your lack of conscience about as physical matter is manifested and the paper that you play in this process, that your life appear like an event that is out of your control does . You can seem you than

You are a victim of the circumstances when in reality you are the cause of those circumstances including the fact you may be experimenting of the riches ó the scarcity of the riches.

The Carnal is the first step of an incredible awakening that you are about to start. Not only you are going to understand of that all that surrounds you but also is constructed you are going to understand how your beliefs and thoughts create the matter, how you harvest what you sow and enclosure how he is than before you ask it for right now you have been given it. This is the way science is itself in putting the day with spirituality and common sense at last and at the same time is explaining them! Think : When somebody tells you that anything is possible under the condition that you believe that he is that way, It Is Not maybe more probable that you create it if you know scientifically, step by step, like doing it and than as a consequence of it the universe change and produce that are you believing ?

One of the benefits to understand the Carnal Cuántica's basic concepts (and only what's basic is what you need to understand) is to be able to see finally in clear form, among other things, how loud slight knowledge like faith and the correct pansy work . Seeing and understanding this knowledge allows having total confidence, to eliminate doubts, to create your reality conciente, beautiful and mightily and, of many manners, turning around but powerful.

The Carnal shows you also like are connected wholes, how all are A very Being that you perpetuate the illusion of individual separated beings.

Also he gives you a peek of as they interact and they connect the Matter with the Spirit and the mind with the Matter, how the creation happens in reality and how we are co creative with God. At the end of this chapter and in the following four chapters, you are going to get surprised of as much as you have power, of how incredible it is the Universe and than easy you are to create any outcome than your choose to .

But first, That it the Carnal Cuántica? The study comes from the basic blocks in which the Universe is constructed. For example, your body is composed of cells. Cells in turn, which are made out of molecules they are compound for atoms that in turn they are made out of subatomic particles such like the electrons. This is the world of the Carnal Cuántica. Everything is made out of big groups of subatomic particles.

Your body, a tree, thoughts, a vehicle, a planet, light and everything else are concentrations of energy. All of them are big sets of approximately the same subatomic particles. The one and only tells apart it is in the way that these particles form a group at each time bigger blocks. The key to create yourself for you and to the world the fact that you surround you is to know as they work .

In order to talk correctly, a subatomic particle is not a particle in the sense in reality in that a sand grain is it. While atoms and big particles are objects, or things, subatomic particles are not that type's objects. Odds are of existence and at the same time multiple existences.

The Quantum Physicist asks that they are these subatomic particles and how they act . Good, subatomic particles are parcels of energy otherwise called quanta. Everything in the universe is made out of energy and those parcels they behave of the most incredible way! ¡They get under our orders! The reason which they would be able to form a group for in the shape of a ship of luxury, for example, is in our individual thought and bus. Do you begin to see the link between the Carnal and the riches? To this point it may be that you had come designing your life at random and unconsciously.

Now you are going to wake up and you are going to do it deliberate and conciente with an address.

Let's begin our journey for the quantum field

That are you have become your body? Of textiles and organs. ¿That are they have become textiles and organs? Of Células. ¿That are they have become Cells? Of Moléculas. ¿That are they have become Molecules? Of atoms. ¿That are they have become atoms? Of subatomic particles. ¿That are they have become subatomic particles? ¿Of Energía? No. They are not done of Energía; They are energy.

You are a great piece of energy to the equal than the rest of everything.

The Spirit and the mind create this energy in physical the way in which you are accustomed to see them.

The Carnal tells us that the act to observe an object causes than the same be there where and when we observed it. The subatomic particles that in turn constitute the atoms are the Energy and finally the matter. This energy exists like waves scattered by the space and the time. Only when an observation is accomplished, the waves become an event localized in space-time, a particle in a time and a place specific. As soon as you withdraw the observation, you become waves again. Therefore, as you see, your observation, your attention to something and you intention, literally create that thing like an event in space-time. This is scientific. We will see like the focus in another chapters, the concentration and the care of attention, the intention and the thought create your reality exactly.

No object is solid. You are made out of parcels of energy that sparkle rapidly. Billion and Trillions Of parcels of energy. They appear and they disappear of the space where the object is . They do not remain at the place.

Then, cause a human body ó a car do solid objects and continuums when in reality we know that they are a field of energy that sparkles rapidly seem? You think about Televisión's image. When you look at a film and you see a passerby walking softly, in reality a roll comes from film that turns to 24 pictures for

Second best, each one lightly different of the following, of way such that your eyes do not detect the breach between them.

Even each one of those pictures is a composition of billion photons sparkling to the velocity of light. That is what is our world, a fast twinkle that causes the illusion to be solid and I continue . Once you understand that your world is done really, really you begin to understand his true behavior and nature. He is there when you change your perspective of him. With your change of perception, you change your world's creation and the first step toward the riches is that one.

All physicists agree in a thing: Subatomic particles, those parcels or as many as, they are not particles in one specific space point and of the time, as a table or a chair are it, but that they are a probability that they exist in several points of the space and of the time. The act of our observation turns them into a particle " phísical"en one specific space and time point and once we withdrew the attention, they become probability again. Imagine that the chair at your house's living room is a great subatomic particle. This would be the way you would behave in: If not you are home and you are not thinking about the chair, this desparecería and you would become the probability of To reappear anywhere in the living room ó, in reality, anywhere in the Universe. When you return to your house and you think about taking a seat in a located chair in living room's specific place and you look for a chair in that place, appear magically!

This is not a history of magic fantasy. iThe subatomic particles behave this way!

What's amazing is that all matter is composed purely of enormous quantities of these particles. Therefore, you perform on all of the matter exactly as a great group would make it out of subatomic particles. The fact that the chair be there stems from our observation of which you are there and deciding that you be there. It is not a totally independent existence.

No type of matter exists completely independent of the observer.

As they say some scientists, if everybody and everything in the universe leave ó to think about her, a physical moon would not be further, would be a probability of existence to face the moon. The act of observation does that probability become a definite thing and converts in void all other possibilities to be in any other place of the same world. The continuous attention maintains it that way, producing the illusion of a solid and continuous existence of the physical moon.

The physiques also uncovered than the quantum particles take decisions. They are propelled for intelligence. Not only that, also they know, instantly, what decisions have taken the other particles anywhere in the universe! This synchronism through the space and of the time is

Instantaneous: Himself they communicate without mediating no time neither crossing no space. In fact, also they move instantly without having that to cross the space ó to take time. They can go from on the spot.

Point B without having that to cross the space between means and the points To and B they can be in different times. Remember the quantum particles are not particles in the sense that usually they devote themselves to to him the word when one thinks about a particle. They are not things that are in a particular place and in one
Instant, they are spread about the time and the space.

Then, is Cual the intelligence that propels them? Good, the mind comes from the Source, God, All That He Is, constituted proportionally of your own individual mind and as well as of the minds " place mats " of the rest of the universe, depending on the theme, matter, vision and fortress of will.

Think carefully on this theme. Think on the fact that all that your holes see is made of these incredible particles that are under your control knob. You consider on the scientific facts that they try now that you are in the cause ó causes once what you surround you was shared, for everything, the fact that you swim of what you observe can exist without your observation. All that you need to do is to elect what you want to observe, to elect it with certainty and consistency and this will cause that the field of energy materialize in that thing in the time, depending on your clarity, focus and certainty. The scientists have discovered than still in the more strict experiments of double isolation, his expectations still influenced the results, concluding that it is impossible to accomplish an experiment where the beech influences zero for part of the observer.

The quantum parcels ó they define particles better like odds of existence. For example, let's say that you have a quantum so-called parcel Mr. X.

Before you ask to talk with the mister X, you do not exist like person. You exist like a potential person. The Mr X is all over the world at the same time, with the variable potential to appear in person in Moscow, New York, Kabul, Tokyo, Sidney, Cape Town ó any other city in the world. Now then, when you call him for his name, you appear where you called him and probability that you appear at any other city becomes zero . After you finish chatted with him, disappear and you stop being a localized person, it spills out like a wave and the probability to appear in any place of the world takes place again. It is the way that Mr would put up with himself a quantum so-called package in X. Bear in mind that everything in this universe is made out of quantum parcels.

Another them property as many as it is that they are multidimensional. Therefore in the previous example, when Mr X is a probability, it is multidimensional. When you locate yourself when calling him for his name, you become a thing in a world of four dimensions (our world just as we know it is, in reality, tetradimensional being its dimensions the length, the width, the height and the time). This is scientific. Now you can see scientifically that our universe is multidimensional in spite of the fact that our very senses be able to detect only length, width, height and time as if they be the only dimensions. Regardless of that, our minds are also multidimensional. Listen to your soul, to your sentiments.

The physical world is made literally of ideas and energy.

In the event that you sense that you do not have somebody can this considers : Einstein and another quantum scientists have proven that all physical matter is made out of parcels of energy that do not have limit in the space in the time and.

This field of energy does not have clear-cut limits. The universe is your body once neither time was extended of space limitless literally. The science has proven also that the mind does not have limits. All minds are
Connected to a very mental field. You are bigger powerful so that you think . Then, you stop sweating for the smallnesses.

You right now have everything. It has been said than before you ask it for right now you have been given it. The science is beginning to prove that, through the quantum physics, this is scientifically true. The infinite intelligence and potentiality to quantum level, the level that you constitute all that surrounds us, and our inherent ability to act on this field is what you provide us to have everything. We are beginning to understand this in a bigger scale so much scientific like spiritually.

You right now have all of the riches beyond your most fantastic dreams. You have it. It may be that you not be living it at this moment but you have it.

Having and experiencing music different things. An easy form to explain it is that you have the capacity to pilot an airplane, mounting waves ó diving but it may be that you have not experienced this aspect of our.

There is nothing to for you to have to make somebody have this ability; You right now are in yours. They right now have done you.

All that you need to do is to experience this ability.

In our life, simply move you raise her consciousness to experience ours aspects same that we always have had and enclosure some that not even imagined that they exist . The quantum field can form an infinite quantity of forms and experiences as from that.

Still further, you right now have done it that way. This book's page is only an one belonging to those things, the word that you are reading is only an one belonging to those things, one of those things is only the next thought that you are going to have but still that way you never may have forecast than would be experimenting these pages. However, your desire to find them, the fact that they appear in your hands has caused . In truth, they always existed . You do not need to predict exactly as they go to work the things, all that you have to do is to desire, having the intention and knowing that it is possible and everything will get fixed in order that you be the host for .

Many physiques that run with subatomic particles are uncovering several interesting things about the universe. For example, they have found that two particles separated by the time and the space can be invisibly linked among themselves and acting in synchrony. They have also found out that the world we lived in seems to be constructed as if you know yourself . This seems to have happened cutting the one total.

The least two statuses, with a status designed to see and the other one to be obvious. The one that was designed for seeing gets under the illusion of the separation it was of designed equal obvious being. It is a necessary and persistent illusion but one is everything in reality.

All universe is made with participation and observation of those that they participate and they observe . The riches are manifested because you and I are examining it. Your certainty, your faith and your attention is that they create it.

In reality, you right now exist like a wave of probability, but you, causes that a thing be definite, an event in one space point and of the time. This goes still more deep. You right now exist like an event but the perception of the time comes up with her separate and separated. Once you understand what the time is and as it works, you will be able to manifest faster and in large numbers.

Now let's go to something a little bit more complicated. We have seen that subatomic particles exist like odds and than when we observed them they cause his localization in one particular space point and of the time. In other words, a particle has the potential to be in the places To, B, C and D. When we observed the place C, the C appears at the place and stops having the possibility to appear in the places To, B and D until we stopped observing the place C. Very well, a new school of thought is going into Everett Wheeler Graham's theory that it says than them as many as they will locate themselves in the four points but in different worlds that they coexist.

In other words, all possibilities are shown at what's physical really but in parallel worlds! There are proofs of this in the investigations accomplished by physiques that they look for signs that the universe is a giant-sized multidimensional hologram. ¿How does this work? Good, when a particle has the chance to be in the places To, B, C and a place to be rather will not elect D only you will choose to be in the four but in order that this be possible, the universe himself the party in four, unless each one of them surrender tells three of the other ones.

In quantum mechanics, calls this interpretation of the multiple worlds ó parallel universes.

Sound wild but think a little. It is definitively possible – nothing is impossible for The Source, God. Many religions tell us than still before we ask it for right now we have been given it. Also they tell us that all that may exist exists Now. Nowadays also we know that the universe splits better ó itself still the fact that the illusion of the separation of way, creates such that a piece may be the observed piece and the other one may be the observer to know oneself to herself.
The Uno breaks to oneself so that you may know oneself and having something with comparing . When one is lonely you have nothing with it as comparing to know what he is .

Your being, your spirit ó nourishing, she is eternal and you exist through the space and the time. Now then, next time decision that you take will split the universe.

You will become conciente of the part of the universe that you elect . You will exist in the other part that you did not elect, but you will not be awake in her, in spite of the fact that still you receive his essences and help to you to know what you elected and vice versa.

Someone else that elected the other world that you did not elect will be
You wake up in him but no here, in your world. Now you can see like the free will it works all the time without going at odds in with oneself and besides, as it is that a truth can be apparently contradictory and regardless of being totally true.

The universe also himself, depart in your present being, your last beings and your future beings. You are in alert those beings to one at the same time (your present being) but your future beings, for example, are able to notice your present being of things which right now they happened for and they pleased us of way of than your not happen for the same reason. It is a very compound and constantly changing womb. All womb moves around with each decision. The life is constituted of eternal processes that they exist at the same time with all the possible existences, at this moment. You, only you elect one to be alert in instant in instant. The physics recently is beginning to try it. On purpose, dreaming is another status of conscience simply.
When you dream you are in another command, lost to the world ó once upon a time.
Can you understand now from where the dreams come ?

Everything this is going to have more sense when you understand what the time is and as it works .

Now we know that the universe came from a similar existence to give it a subatomic particle. Ever since one has been expanding a higher velocity of the one that you may imagine. They form oceans, worlds and similar things, everything by courtesy of the Carnal Cuántica take shape .

But this is still more magic. The universe is giving continuously to birth new universes. Many physiques proofs of behavior in agreement see now the theory of the parallel universes and of several interpretations of the Carnal Cuántica that exist for that purpose. They create the majority of the physical that ours is a continuous existence but chaotic risky ó, principally because they have not found another reason for existence. But to you: What paper does the Spirit play? What role do you obey you, your being, your soul? ¿It may be that do your elections be the cause of this chaotic fragmentation of the worlds? like to the physiques exclude of his investigations to the spirit in spite of the fact that this gives birth to the matter and upside-down no. Think a little. Einstein is one of the few physiques that you were unable to operate to believe than everything happens of chaotic form ó risky. You said that he stood against being believed for God to play dice

Find out about this: The quantum world is the real world. The world that you see with your eyes is an imperfect perception of a group of quantum activities simply. Still that way, the quantum activity results from you you are the First Cause. Never you think that what you see with your eyes is the First Cause you are only an effect. A good illustration thereby is the famous experiment of Schroedinger's cat.

(proposed for Erwin Schroedinger, whom he was granted the reward Física's Nobel in 1933). What happens when you put yourself in a cat a box with a bottle of poison gas shows . A device within the box is placed, which can open the gas bottle and therefore killing the cat. Next one closes the box of way that it may be failed to see that there is inside. On purpose, the bottle's gas only becomes poisonous if a radioactive atom in particular, it breaks down.

You show the Carnal Cuántica than this radioactive atom decay exists simultaneously in the statuses of decay and of no until now in that you try on it (that is until you open the box to see if the cat is alive or dead person). (everything exists in all of the possible statuses until you observe it) You Can Not Know if the cat the dead unopened ó is alive the box. While the box remain closed, the cat is, at the same time, so much alive like dead person.

For this reason it is that the Carnal Cuántica is so crazy two contradictory statuses exist at the same time! He is, the moment that one opens the box to verify if the radioactive atom broke down ó itself not and therefore if the cat the dead ó, when he materializes one of these two statuses itself is alive .

Still that way, the physicists know now than (1) his expectations and thoughts have influence in the results and (2) the interpretation of the multiple worlds ó parallel universes you tell us than both happen but the world himself part in both and we took, in some level, that that we elected .

We know now that everything in the universe is a duality wave particle. What this means is everything, once your body was included and your car, they are a wave and a particle at the same time.

There is no difference enter you and light – except for the fact that you have the light one different wavelength of yours but you are the same thing as light approximately.

This the Physics says it at present. Thousands of years get back, several spiritual teachers did not say it we come from light.

You are light. In fact, if you examine your body with a microscope you would find that over the 99 % is space.

The rest, the part solidifies, is only a set of exactly the same things of which light, of the same subatomic particles is compound. In reality, even the space is full of energy.

Your mind, as from the spirit, holds together the body like a solid unit. Your mind does the same thing with something else that they surround you. All matter is created using information of the mind of the minds of those that they surround you and and of the rest of the universe.

The equation of Einstein And mc establishes that the energy contained in any matter is equal to the mass of that matter multiplied by the square of the velocity of light (that is a very large number!). This shows you than 1) Still the smaller parts of the matter have a dreadful quantity of energy within himself (that way he is like work the nuclear explosions) and 2) Thou and the rest of the things very music energy assembled with information of the mind.

The subatomic world is not for static nothing. An incredible dance is of unending creation and destruction,
With particles that they get destroyed they conceive new particles to themselves and in that same destruction.

The heft of the particles they have an unimaginable short life (thousand millionth in the second position). The complete universe is entertaining oneself itself constantly. You can imagine what it means that everything is swept, cleaned and reset solo a little bit different second ó's each trillonésima something like that. Very well, here there is incredible something else: When a particle is made, travel instantly to the velocity of light!. Literally, we come from light just as they tell it many stories of the creation. A thing more: The particles can move forward or to the back in the time. ¡This is the substance of you are made and that which be left over of you have control!

A thing like the vacuum does not exist . All space is full of energy – the same energy of whom you are made and the rest of the things. The case is that your five senses of sight, touch, ear, sense of smell and pleasure do not detect the a great variety of other types so that they exist in the universe. In other words, only you can perceive forms that are detectable with your five senses (to the least that you had developed another senses). This does not mean for those detectable forms for the human beings, physical forms to be the only things that exist in the universe. You think about the universe like a hologram. Anyway, this dreams it up of everything you are to know that you are a part of a giant-sized ocean of energy and that literally nothing.

You separate you from nothing. The only separation that you perceive is an illusion caused by the limitations of the five senses. Literally we are One.

We are the organic whole whose parts are changing all the time.

Each part can face the other parts and each part has his own level of conscience and knowledge. Still that way, you perform on the whole exactly like the whole, while parties act as part of the whole with individual ownerships and with properties altogether.

The Dr. J.S. Bell, a physique of the CERN (European Organization for the Nuclear Investigation) in Switzerland, you found a mathematical check in 1964 in the one that you were showing than all the separated parts the universe they are connected of an immediate and close form. Many experiments evidence than the particles separated by the space and the time know somehow what the other one is doing at the exact moment in the that the other particle is doing it. In other words, they do not get in touch . The communication takes time and requires a message. This is different . They know without having that to get in touch. They act simultaneously like if they be connected intimately somehow through the space and the time, of a way such that they are not affected by the separation of space and time between them.

The fact that you show Bell's mathematical construction is that the action of a subatomic particle depends on something that is succeeding someplace else another subatomic particle after very deep thing. In other words, all subatomic events are effects.

Of other ones and causes of a similar amounts. ¡This puts under a completely different light the law of cause and effect, of karma and to harvest what you sow ! The cause-and-effect laws, karma, you did not sole spiritual but also she is scientific.

We have come saying that all thoughts of the universe, place mats and buses cause than energy himself, configure in the physical reality that we experienced . This is that way, but more fortress has a cause.

This is the being – the status of being. There are many statuses of the being such like to be happy, being fast, being rich, etc. This is the loudest cause, the First Cause of everything. This is that way owed to that it is Espíritu's declaration, of the you per se. They generate thoughts of their own of that status as from a status.

Here there is another way to understanding as it is that all are one: The Science evidences than we same and everything are done of energy and interchange energy with everything them besides things, all the time and of a very complex way. Energy is the building block of all the matter. The same energy of which your meat is compound is the same of which the bricks of your house and the trees that are out are compound. There are no energy of the tree and the man's energy. The same energy is everything. The energy constantly is flowing, changing form all the time. This is a very explanation simplified of an enough compound thing.

Level quantum, everything looks like a great stone basin of soup of energy, an ocean of energy that is always flowing, an ocean that has different concentrations and essences in several points. You imagine an ocean with a hot zone, a turbulent zone and and so on. (the ocean represent the soup of energy and zones represent the several physical objects such like your body ó a tree).

The hot zone interchanges molecules with the rest of the ocean. The turbulent zone also you interchange water molecules with the rest of the ocean but the essence of that hot zone remains hot and give it the turbulent zone remains as such. There are interchange and flow to molecular level but the zones continue to be hot ó in a most elevated level turbulent in spite of the fact that the molecules that were fixing her a moment ago right now have gone away and they have replaced themselves for another one of another regions. The molecules at the hot zone change but the essence of the characteristics of that zone remains hot. The characteristics of a region can remain equal in spite of the fact that the particles that form that region are different always, entering and leaving of and toward another regions. This is the way that we are quantum level, a great field of energy interconnected with locations of characteristics in. We shared the same energy with all the rest and we assumed only characteristics regardless of that. It is a very compound womb, a compound cobweb.

Now let's put it a little more complicated bit than the example of the ocean since this is of no use for us for our explanation. Imagine now that there are two people at a room. Both are depressed and sad.

His energy levels are basses. Join theirs you tell a joke and the other one laughs . The person that told the joke causes than her, someone else that began to laugh raise your energy level and turn around more vibrating. This causes than the one that you said the joke than also laugh because there is a new joy on the air and they are sharing a joke. The person To the B caused a change in the person and this returned to cause a change in To. ¿Have you seen as you feel good after you tell jokes to a great many people and better than if you tell it to a very person? That people take the joke that you told and is told to it to another friends and these they tell it to his friends's friends and it spills out . Good, the Universe is much more complicated than that. A change in one part of that giant-sized field of energy spreads out and causes changes in the next parts to him and these cause changes at the environs to themselves and the effect is continued forever! ¿Can you imagine that? ¡Your smile changes the composition of all the universe!

Scientifically! Your irritation also does the same thing.

Any thing that you make, any thought that you have, spreads out forever and the composition of all the universe no matter that the change be so little changes .

Now here something that is still more interesting goes . Because you are a part of that universe, that propagation, like a curl, drives you you giving you a touch of a similar essence. You cause a change at the field of energy that you surround you and this propagates itself touching everything and of course like reaction.

Everything goes back to you multiplied! This is similar to what happens with the curls that are formed like growing circles when you put the finger in a glass with water. But, in our case those waves keep on forever and also they collide with another locations of energy, they cause changes in them, these react, in a manner of speaking, they send his own waves and these bounced waves touch you, change you and you react and this incredible dance keeps on forever. This is the way in than, scientifically, works the law of cause and effect and makes it out of multiplicative way. This succeeds level of energy and level spiritual. Both times, the improvement of an individual in the system causes an improvement of all the system and an improvement of all the system causes an improvement in the individual. The inverse also is true .

Supposing that we leave behind constants everything them besides things, an individual and a society will manage to have riches and happiness insofar as they understand that the universe is made and they learn to control the creation.

It is our brief introduction we have become of whom to substance.

Do not get worried if something does not make sense – it is going to make sense later on, as you keep on reading and find another themes that relate to each other to him. You need not a Carnal Cuántica's teacher in order to be rich. It is sufficient with this chapter. The objective of the same era simply showing up that your world and beginning to show up is done really that you get under your control.

The rest of this book will show you like to exercise the control and reading the return that you give you.

The Carnal proves to that the world is not the hard and unchangeable thing as you feign to be. Instead of it, ó is a very fluid place that is built continuously using our individual thoughts and buses and statuses of the being like society, country, family, planet, solar system universe. We have begun to take the top off the illusion. We know of that our three-dimensional experiences are done and we have begun to examine as it is that we manufactured them.

The following step is to know as the other party is world. ¿That is the fourth dimension in reality, that is, the time? Time becomes of examining the time. Let's be going to that

The Truth On the Time: The time does not exist to less than that you say the opposite.

The time is a funny thing. Full of fun. The bigger trick the fact that you had played the time is to make believe that us he exists and that we are under our control. Instead of it, the time is a total illusion, a loud and persistent illusion.

Than wonderful news! The time is an illusion created by you. Once you understand as you create this illusion, you begin to entertain your pleasure, in form conciente and deliberate instead of doing it unconsciously and in shape accidental as you may have been doing it until now.

What is the time? That do I have to see in the course of time? ¿How would I have to think and being in respect of the time so that you may experience riches and something else in large numbers and more rapidly? This chapter will begin to answer those questions. These concepts will become more real in subsequent chapters.

The only time that really exists is the Ahora

The distinction between past, present and future is only an illusion, however, persistent. - Albert Einstein

The fact that the time be short long ó, that the space be broad or narrow, depends on the mind.
Whose minds are idle for those they can sense that a day lasts for a millennium and for those whose thought is expansive they can perceive an eat small house so roomy like the universe. Hung Tzu ch'eng

The time flows in all directions, not only forward as you show off .
The past, the present and the future exist simultaneously.

A simple explanation goes from what the time is here. It is an explanation outrageously simplified but serve for the moment.

Imagine a football field with 10 objects scattered around.

Now, imagine than a determined object To represent a little newborn boy and another object you represent B to a little boy of 10 elderly years. If the object To travel toward the object B, what you now call 10 years time would take the journey.

This is 10 years of human time in a little boy's growth. Now it goes down a little bit but complicated: What would happen if did the football field contract? The object To the object would catch up with B and it would go by for all the experiences of 10 years of infancy, but the sensation of the time would change .

In other words, they would experience 10 years of a very different way.

If the same field of soccer contract enough, ten years would be able to seem like an instant. You have experienced this often. When you are passing her well you seem you than the time, fly . You do not indicate that they spend the time ; Still that way your clock says that hours happened because your clock is designed in order that you take the same amount from time in order that the second hand of the clock move between two consecutive check marks. But you are not laid plans that way. The time is the movement to your conscience suffering through the preexisting events in the continuous space-time. Right away you will see that it is what this means .
The field of life we inhabited in is not static, you change his dimensions continuously. For that reason it is that we have all of the time that to readjust clocks worldwide for that this madness called time make sense, but this is that way because we thought about the time like periods of constant slices.

Not so. Our misinterpretation of conscience moving of a preexisting event at the field of the Life to the following, is that one as you will be able to see right away.

The life's field is not static and conscience does not move a constant gear itself. The field is able not to change so fast for us and our conscience is able not to change its so fast velocity, to the least that we that way want it and the fact that usually we did not see is that's why so much those differences and es perceived that the time is unchanging.

But, if you travel very fast by a spaceship, as you have for sure heard somebody mentioning in Einstein's Theory of Relativity, you would be able to do that the time spend more DSPACE and even you would be able to travel backwards in the time.

The time not only that a sensation of events that they pass and at the most fast more slow ó pass events, all the more fast ó but the calibration of the span will be slow. What you change is nothing but take as much time as the calibration of the interval (a minute no longer take a minute).

Agreed, let's go back to the football field. Imagine that you be one of the objects. You would feel the time as you move around the field, passing near another objects it not true? Yes. Now imagine that you were born moving faster, let's say, three times velocity. The time would seem shorter. ¡Now imagine that you are the football field! Or still further, an object the sufficiently big to cover all of the field. ¡Now indeed we are.

The time would stop existing for you.
Because you are the field and you can feel, being ringing and being all of the objects in yourself at the same time, forever, the journey from an object toward another object would not exist . In that case everything would be happening Here, Now. Everything.

The 10 objects would be succeeding the same time for you, always. This is the Ahora's eternal moment, Aquí. Everything that can happen what can come into being, the past, the present and the future, all are at a run to the same time at an immense field in the universe, everything. Your conscience and knowledge are awake only to a little zone of the field in any moment and as you move around passing near those objects from one point to the other one, you experiment time like a sensation of past, present and future. The field per se does not experience time; Only you experience an eternal process that is always succeeding Now, Here, Always. You can imagine to all the field like The Source.

As you expand your conscience and knowledge, as you take more and more from zones the field, they contract the time to you. ¿Can you see it?

Now what's incredible one belonging to this is the mind and Being (our soul, spirit ó one way or another than your call it) is much more big than the physical body.

We use to think about the soul ó the Being like a little thing contained within the body. That is only human thought relating things to containers.

Possibility of than the soul, being but powerful than the body, in reality maintain once the body was joined and surround it? And that do the mind contain the brain and to the nervous system and surround it? If you bear in mind that the soul and the mind are major than the body and the brain, have You Thought occasionally where they finish ? how many meters of your body? ¿Or maybe will he be to how many kilometers? ¿Or perhaps how many light years of your body to does your soul finish? Would it not be impossible that your Mente and your Alma be thousands of million times as much big than your body (Why not?) They are infinite and eternal.

Regardless of that, this giant-sized and powerful Ser, you are . Anyway, let's revert to the theme of the riches. If you wish to a lot of riches experiment and rapidly, you need to understand clearly that it is the time, as it works and I eat to take control of the same. Also you need to understand your Being and your Ser's composition relating to everything else in addition to the above, either physique ó not physique. Everything is a subject of expanded conscience, correct status, correct thought and election. These drive to that your conscience wake up the delicious parts altogether and do it in each time bigger circles.

The now it is the only moment that exists . The Ahora's eternal moment is all that exists . You will be able to remember the past and dreaming of the future but you soled you can be and existing Here, Now. ¡Contract an irrevocable commitment with yourself to make of the Now the best moment of your life!

Not occupy you over the past neither living persons in the future. Your only moment is the now. Live in the Now.

As you are going to check right away, a mirror is your world outside to your inner world. You will check in this book as it is that this is that way.

Do you feel as if you have not enough time to do what you want ? The people that has a short time in what's outside are short of time in the inside. They act, they think and they talk believing that they are short of time. You stop thinking and saying that you have not enough time . Do not believe it not even for single moment. The Universo does not have scarcity of anything, even of time and you neither have scarcity, except that one than yourself you construct yourself. Believing in any type of scarcity does the fact that your conscience be drive smaller and more slow in order to that you may live what you are believing .

The present moment is the bigger gift that you may have. It is created perfectly for you according to your declaration of desires. You declare the desires through the thoughts that before you have held like the truest, the statuses of being, the words and actions. The present is something than your you send off yourself, a perfect present moment. You allow experimenting, to savor, to make a revision and to change your last thoughts, statuses of being, words and actions. I know thanked for the present since you know that you can change it since you allow experiencing your Being and besides because his complete existence is about to serve you. Swearing, condemning and judging the present only will do that it linger on . Whereat you resist that, you pass judgement and convictions, persist . You embrace what and you bring in the light to examine without judgments, in clear and honest form, you reveal the lessons that you are on the lookout for, the key to the following level that you chase .

The time is only an illusion produced by the succession of our conscientious statuses as we traveled through the eternal duration and you do not exist where the conscience which the illusion can be in produced and that asleep H.P. Blavatsky remains in does not exist

The future acts on the present as much as you do it on the past – Nietzche

The first time that you do something is a journey of discovery. You take details and you learn many new things. At this point, there are no labels neither memoirs to afford to prejudge the new experience. Learning is in its maximum point. The hundredth time the fact that you do it is to very different pocket money.

The repetition brings inconciencia for the greater part of the people. They do the people's majority and the things see that more frequently happen in his lives in an irresponsible and sleeping status. Because they have seen ó once something was done once they overturn toward relying on memory and in the labels that constructed in his minds the first time. The learning and the discovery fall to zero and take control the memory of last experiences. ¿That do you be useful for living today based in your report of the past? ¡You lose the gift of the present moment completely! I work ó in your business, you take maybe an absolutely fresh look to work, co-workers and customers each day you guide ó for the way in than them do you know of the past?

Everything changes and the use of memory makes it impossible to see your change, to see things just as they are really.

Try to forget everything, bring near of that you are looking ; You will discover a brand-new world and you will grow a lot faster, will grow your riches and you being a lot faster.

Think about this. Succeed often that a stranger appreciates in your co-worker ó consort something that you fail to notice totally and this is owed you do not look at them to as if they be brand-new for you. The memory has its show but a great many people abuse of her same and frequently so that they are not for beneficial to them.

You make a decision now to confront each experience as good as new, electing to observe her as if you would never before have it lived. Ó decides not to anticipate a specific appearance behavior, an anticipation based in your memory and emotions. You practice the detachment of the outcome but have certainty of your elections and intentions and you will find a world that has come over to you hiding all the time in front of your own noses.

You elect all the times the happiness, living the present instant and joy. You thank the present moment throughout what you bring you in experiences to have a good time and in the opportunities to see you yourself like eras before and to grow toward a still bigger being.

Bring in a minute present you waking up, your conscience, your thoughts and looking . The life and all opportunities to advance are in the present, the eternal moment of the Here And Now.

No to you pushes all day long forward where you wish to be.

The imagination on the future is very good, since what you use is for creating your future, but the attendee has a great value. Only you can put oneself in the future through performing on and living in the present. Not happen all day long dreaming on a future moment saying you if at least, escaping mentally of your present and engaging in an imaginary tomorrow, exceeding the day in a status dream, being alone half an alert and conciente of the details that happen in your day – in reality they brake these things in your journey to a better tomorrow. As much as it is necessary for you to put goals, is also necessary that you embrace and you experience the present and you perform on conciente in him and giving you account of the same, if it is that you want to progress. Remember, the universe only can utilize the present moment to send you clues, people, events and the imaginary future in your head can not use opportunities to advance. Instead of pursuing a better future pushing your conscience toward the future, bring her at the present time and quit than the future hunt her here.

The time's wheels are mysterious. The time is a concept of the mind. There is no concept of the time without the mind. The mind annihilates . You will surpass the time. You will enter the world without time. You will live in the Eternal. –Sivananda

Not be left to you in the past, do not dream of the future, Buddha Concentrates the mind in the present moment (566-488 a.C.)

Be careful in how you put dates limits. The time does not exist like one absolute. The Carnal Cuántica, our spirituality and our understanding of everything.

They tell us the eternity than the only time than only You exist really it is the Ahora and the only place he is Here. Here, Now. For example, imagine that you have the goal to become millionaire multimillion ó in a year. Think about this: ¿Because did you elect a year? It is a very arbitrary time limit. He is simply like to take a date. It's possible that The Source may create that millionaire or multimillionaire in an instant. Nothing is difficult for The Source, God. Then, because putting a date at random for the fulfillment so that he can be caught up with in the twinkling of an eye ó in no time that it is better off adapted to you and on the one that you did not consider ?

Putting dates limits for you also introduces fear and you doubt (do you be going to be able to do one's job by then?) And many times the one thing that you get is that you rein in you (What Happens If you may have achieved it long time before but does your mind continue looking toward that distant date?). This also causes that you be for difficult to you becoming unattached which is crucial stops than infinity and unpredictable power of organization that The Source has, work in optimal form.

On the other hand, saying someday I am going to be worth millions or multimillion neither you are good; He is still worse. Instead of it, think at present like the only time that exists . Look at it and know it like the only real time. "Now I am worth millions ". Now. I am . The way always you should think in, that to act, to speak and to feel be more than enough anything in is that one.

When the people ask you when, say them soon. Jesús always said soon when somebody asked him when. Soon you become adapted much better.

Here that is to establish a specific date. In your mind, everything is Now, becoming something, he always has been .

Remember, even the Carnal Cuántica proves that the time is not what you consider . When you say Now I am worth millions and the physical evidence that you surround you does not evidence it, does not mean for you to as telling lies to yourself. In reality the present moment contains all of the existence in all the possibilities, everything at the same time, including give it that you are rich . That affirmation is true, the ones that tell lies are eyes. Einstein said once it is not important for the illusion of the time, the illusion of past, present and future to be able to be so persistent continue to be an illusion.

Have patience with the events. If you press ó you force them, you interfere and you delay them. Nature is perfect. If you want results but rapids, the correct way to shorten times is raising your certainty, increasing the clarity of your imagination, being a very mind (no changing idea), concentrating and what's more important, lifting the level of your conscience from the level of the mind conciente to the level of conscious, subconscious, supra-conscious and Being. People's majority they do not become aware of his subconscious and super-conscious identities. If you have perfect conscience, imagination, faith, certainty and clarity you can create results instantly. You are beginning and each time you will improve more as you spend the time - simply I know intentional in these things and you choose to be more awake and true and it will happen . Do not be impatient because you put you that in a state to want and delay still more the results.

I Am . Present tense. This is the way that you should write your goals in. It is useless to write them if not you think about them in the present tense of the I am . Give account of your thoughts and I know intentional with them being assured that you always consider your goals and intentions in the present tense I Am all day long. "I am " it is the order that they devote themselves to to him the universe to start immediately the process to manifest the desires in the physical reality. A declaration is of the status of the being upon theow..

You must understand and remembering this very clearly, always. When you mean to have ó to experience something, you must know that you right now have it.

Really, you right now have it. All that you are going to make out to depart there is to receive it, to take possession of him. In reality, you will be coming to realize something that always has existed within you. At this moment, in the meantime you read this sentence, you right now are very but very rich. From now on, all that you are going to do is to take possession of this riches, to receive it, ó, in order to be more exact, to be awake to she. I know thanked now for the things that you wish to experiment, because you know that you right now have those things. The Gratitud Ahora accelerates the manifestation of what you have elected because you reaffirm your faith and your status of being.

Nothing is for to come neither swims to happen but the eternal Ahora always will last .
- Abraham Cowley (1618-1667) -

The time is a sort of river of events that they pass and the current is loudly; So ready how a thing appears at sight it is swept and something else takes its place and this will also be swept.- I Mark Aurelio (121 180 A.D.)

When you take a seat with one the pretty girl during two hours think that it is only a minute but if you sit on the hot stove during a minute you think that they were two hours. That is relativity. –Albert Einstein

There are a bridge between the Time and the Eternity; This bridge is the man's Spirit. Neither the day neither the night cross the bridge, neither advanced age, neither death nor grief. – Upanishads

Know the value of the time; Capture it, make good use of it and you enjoy each moment of him. Swim of leisure, swim of laziness, swim from quitting for morning: Never leave stops morning you can do that today. – Chesterfield (1694-1773) -

You wake up in the morning and Ah! You the wallet gets dirty with with twenty four hours magical textile of the universe of your life magically. Nobody can take it out to you.

Nobody receives more ó less than what you receive . Waste valued raw material yourly infinitely as you feel it like having and they never are going to deny the offer to you. Furthermore, you never can borrow of the future. It is impossible to fall into debt. – Arnold Bennett-

Eternity is the infinite existence of each instant of time. If we conceived the time like a line, then this line will be run through in each point by eternity's lines. Each point of the line of the time will be a line in eternity. The time's line will be an eternity's diagram. The Eternidad has a dimension more than the time. - Gurdjieff (1873-1949) -

That piece of nonsense is to be afraid of the thought to throw the life by a single stroke and not to have inconvenient to throw it in pieces and bread crumbs. - John Howe

The attendee is a mathematical line that divides that part of eternal duration which we called future of that one the one that we called for happened to.
H. P. Blavatsky

According to Albert Einstein and another scientists, the time and the space do not exist like separate entities. The time is not a separated thing and the space (fact of length, width and height) neither is another matter separated.

On the contrary, they are a very thing, a continuum once the continuous space-time was called. You try to understand this and to use it. You not believe in the time like something once you was separated from by what you have to pass as it is dictated. Else you take time to understand the time, your pansy is going to direct, your plans, your system of beliefs and your experiences.

The time is not a straight line which we have to walk .

Imagine like a net of low tunnels the city. In order to go from the house To home I join B you can use anyone of the many routes, some more long and shorter others. For example, in 1930 you may have taken 30 years a person to make a fortune by hard work. Now ó can drink two years less. We have increased our conscience of the riches of certainty and and this allows taking a shorter path.

The continuous space-time is like a sheet of paper that contains all of the possible events another way to thinking. Each one of the possible events, everything, you are in the paper. After imagine to to you be a pencil with an eye.

Next you imagine that that piece of paper wraps up the pencil. Then now you are the pencil and are completely wrapped up in the paper that contains all of the possible events; But only an eye of way that only you can see the event that is happening in the piece of paper at the region where the eye is has this pencil. The pencil's eye can travel up, stops I get down and around the pencil. It is a movable eye that can get anywhere of the surface of the pencil. At the most quickly you move the eye, more events you can Seeing . Now imagine that you can increase the size of the eye. At the most big the eye is made, you can see more events simultaneously. Very well, the last thing: Imagine that you can multiply the quantity of eyes on you, the pencil.

When the pencil has more eyes, go but life, but events simultaneously in place of of one accord for time.

When you increase the quantity of eyes and the size of each eye, you see more in less time. Well then, let's replace the word eye in all this explanation with the word conscience. An increase in the size and quantity of eyes is equivalent to an increase of conscience, in being awake to all of that exists . Therefrom our evolution is treated like human beings, increasing the conscience. In accord, and then, What is the pencil?

Your Ser, your spirit, your soul ó like you want to call her. You are a multidimensional and eternal being, fact to the image and God's similarity, The Source.

As you grow, you increase the conscience and the awakening of this done.

Now, imagine that your best friend is another pencil with the same piece of paper wrapped up around. Now you the pencils with eyes and both are two they are wrapped up in his respective pieces of paper. When your eyes himself they find they experience the same fragment altogether that Is at the same time. You two decide to meet, simply electing to meet. In that continuum, you can decide to find with whatever you come up with it. There are no obstacles and the free will, insofar as you exercise it with certainty operates without error. The knowledge of this, you are to can. Joy is to be able to use this knowledge to obtain riches and any other desire that you have in my whole life.

In this way they can understand interactions. In truth, you, your friend and everybody are individuaciones of the same pencil of such way.

When (the conscience) they find your eyes, they experience being together, but they always are close to a most elevated level.

The events that they seem you simultaneous they can happen to different hour for another observer, depending on his relative movement. Imagine that you are within a big box in a truck that moves around of south to north.

There is a lamp that catches fire and you put out periodically in the center of the box. Granted that you are in the room time that light gets to the north wall to the same time than to the southern wall. Enclosure you can measure and you find that really he is that way. After you verify that light gets simultaneously to all of walls. This is going to be your truth.

Now you suppose that a woman has the side of the route and that your box has a glass window. The woman can see toward inside your box through the window. Because she is static and you and your lamp are on the go, the experience of her is going to be different. She sees that the north wall receives the light lightly after than the southern wall because the north wall is separated of the source of light while the southern wall is moving toward the same (remember the box moves of south to north and the lamp is downtown). The woman can measure this experience and verifying that he is that way. She verifies that the southern wall receives the light rather than the north wall and his truth will be that one. As you can see, two contradictory truths, but exact, they can coexist. How can he be?

For the universe, these are only events. The time is local for the observer, that is, you. In reality, there is no time, only events all at the same time. What you do that you feel as if the time exist is your movement through those events. It is your expansion of conscience that causes that you move faster through more events. An expanded conscience allows you grabbing more events for instant. An increase in the conscience of the riches does that you make a fortune more rapidly, that you move to crosswise extra events more rapidly, that you see each event more rapidly. In reality this drives when being your conciente and experiencing over the Whole that It Is for instant, in such a way that he seem as if you have more things and as a consequence it appears as if you have more riches.

What you do that you experience more for anytime of the Ahora is an increase in the quantity of correct decisions that you take as well as an increase in the quantity of desires and objectives expressed with certainty. Increasing your goals, your mental imagery and maintaining them firm and true, you see more riches. This is one of the secrets of the time.

Hermann Minkowski, the professor of mathematics of Einstein, you elaborated a set of equations that they try than everything the past, attendee and a person's future find themselves in an only point, The Ahora and in an only place, Aquí (anywhere that the individual be doing observations).

All events in the universe are happening simultaneously in the Ahora's this eternal moment.

The illusion that we have they happen of of of one accord at the same time and ó raises 's consciousness in a sequence that seems to flow, is for the kind of perception that we have in this human form. This causes that we see very narrow sections of the continuous space-time for time. We see a little slice, next her following, next her following and that way all the time.

But we can expand or contracting the slices to capture a lot more ó much less for time. A person with conscience of riches has a sight once highly what He Is was lifted of Everything and therefore an experience much more expanded.

Everything that for us constitutes the past, the present and the future is given at a block Each observer, as its time happens, discover, in a manner of speaking, news cut in slices of space time which appear to him like successive worldly material aspects, to sorrow of than in reality, the layout of the events that compose the space-time they exist rather than his knowledge of them.- Broglie's Louis, Reward Nobel of Physics in 1929.

The time only exists in your mind. Your mind often wants to ó in the memory of the past live in anticipation of the future. This is what to a large extent the psychological time creates . Another time is that one. There are many forms of illusory time and the psychological time is one theirs. You experience this time's type when your mind this in anticipation of something in the future ó in the memory of the past. This waiting and remembering creates time, a lot of pain and tension. This is unnecessary. The most fruitful that you can do is not to remember ó not to wait but to observe, to experiment and creating Now.

Creating Now. Observing, it is eternal and true of the true nature of the universe to experiment and to create Now.

Everything happens Now. You remember the past Now. You dream your future Now. You learn from the past Now. When you were really in the past, still era the Ahora (are you in that point of the past, if Qué would have asked somebody doing now?, You would have indicated that still it was the Ahora). You work toward your future, Now. You will come to your future in the Ahora.

You will live in your future, Now. You always are Here, Now. You can not be in no place else. Being, the being it happens only Now. There is nothing to for you to be able to do in any other moment except in the Ahora. Attempt it. Make somewhat yesterday ó tomorrow at this moment. ¡Impossibility! Solo you can be and Doing in the Ahora. Everything happens Now. Even tomorrow it is happening Now, he is Now. ¿Can you see like works eternity? You can not escape. Trying to escape of the Ahora in the mind, it is totally useless and painful. You are like trying had it not been where He Is everything. For that reason Buda, Jesús and many other teachers taught us to not to worry about the future. They taught us how to be still, to be present, to be awake, to enjoy the Ahora, to live a moment at the same time and to be concientes.

The time is a creation of the focused mind principally toward to the past and toward an imaginary future that is out of the Ahora. An easy form to understand it is to go into the phenomenon of the dream. When you wake up, you feel as if recently you were going to fall asleep in spite of the fact that you passed eight hours in bed.

They do not feel those eight hours of the same way what if you would have been alert in during the same amount hours. As you will know about for reports investigations, the people dream almost all night long and remembers only a few dreams ó no one. As you can see, he is her I did not mention, her no memory and the no escapism toward the future that makes you feel like if sleeping go almost eternal. When you sleep, your mind and your soul are immediate in the same place, and that place, it is always the Ahora.

Maintaining constants them besides things, an individual ó society will achieve the riches and happiness inasmuch as they take place counts of the illusion of the time, teach it and use her of the correct way.

The time directly is related to conscience and this, it is what life makes to experience to you. There is an obvious and direct link. Now you know it and therefore, you are much more awake. This finding out will begin to lead you to the increase of conscience. This, in turn, will bring you near still more toward the riches. It may be that in the beginning you not see it, but insofar as you maintain your intention to be conciente of the riches firmly and be I come to realize it, it will happen . Perhaps not know exactly when it will begin to happen. In reality right now it has begun to happen. You will see than you become more awake with each moment that happens until you will one day look in hindsight with running of the days and you will check as much as you have changed .
Do not get worried if not you have understood fully the illusion of the time. Continue reading and the subsequent chapters will clarify your ideas. Always there will be plus for knowing ; The capes never finish.

Measure that you discover over they and deep truths, you will enjoy life and you will turn her into an exciting journey. Regardless of that, keep on always balanced.

Now that we have the four grown-over dimensions and we know of that our experience is constructed here below, we can examine what constructs her, that causes this construction. We already gave a look to the materials and now we will examine the constructor. First we will see like the constructor construct using the status of being, the thought, words, the actions and the laws of the universe. After we will examine the constructor.

Who is the constructor? I give you a clue: Yourself. All beings are construction united constructors.

Before examining the constructor, we will see first like the constructor construct with the space materials and the time. ¿How do you construct?

Specifically, as would you construct a happy and very rich life? Let's see

Imagery of the mind: The working drawings of life.

The following step to make a fortune is to know how the life works . In other words, How is the fact that the force of life receives instructions and turns them into things and new moments as from the field of quantum energy? ¿What type of instructions receives her
Life of form such that you may create each new.

In-fashion moment? In than format and specifications?

It looks for this chapter answering these questions with your help. They are the most basic questions, but they must be responded before more delve deeply into the theme of the essence of the riches. In reality, you right now should have listened to these beginnings previously, they are not nothing new but very not much people understand them in depth and still the littles that are understood, do not use themselves, regardless of his simplicity.

In spite of the fact that imagination constitutes a so basic part to make a fortune, an integral part is of that process. All the other ones parts use this in any stage. The imagery of your mind are, literally, the diagrams of engineering as from which you forge for yourself your world.

In Carnal Cuántica's chapter, you can see scientifically as the reality is produced. You will see also the interconnections of the imagination with the Life in the rest of the chapters. But let's see the imagination, the imagery of the mind for themselves first.

The imagery of the mind are the life, expressed

The imagery of the mind are the Life, expressed. What this means is Life, The Source, use your thoughts, your mental imagery like the instructions with which creating your reality in the world material. The Life expresses your mental imagery in physical reality. To express is to make somebody know, to bear witness, to communicate, to articulate and to transmit. The force of Life makes known the thoughts of yourself and give them the other ones forming them in terms of experiences and objects.

The physical world. You experience your of one's own firsthand thoughts, the imagery of your mind, in such a way that you know as they are adequate and as no. That way it is as you know yourself to yourself, that way it is as you experiment to yourself and that way you are like increase. This is the supreme worldly physical purpose we are in. It is designed to afford to experience an idea, his effects and consequences.

The life does not select which ones of your imagery expressing and which ones not.

How would you elect them to you? Therefore, expresses them to everything inasmuch as you have them and you believe in them. In truth, you have free free will. This is really free due to the fact that really you obtain results all time long, not only sometimes and obtains them exactly. Later we will see like works exactly the free will, in spite of the fact that look like impossible I may do it that way. Imagination is the force that leads you to spaces where you never have been . Napoleón Hill said that imagination is the more wonderful miraculous and inconceivably powerful force that the world had known .

Life is imagination, imagery of the mind expressed in physical form.

Feed your imagination daily and precisely with imagery, films and concentration. Use time doing this in detail every day. Imagination is the most powerful force and he is that way because Life uses them for knowledge that it is what you have to create from now on. They pass the people's heft.

Very little time imagining . They imagine using pointless thoughts, at random and after ask themselves cause his lives are not rich. The imagination has power to construct or to destroy your life. She is your mind and your election.

Imagery, Imagery, Imagery! Imagery of the mind are the life, expressed. Your life just as you desire that he be, imagines perfect image, with colors and details, in your mind, every day. Pass an hour per day constructing your imagery and during the rest of the day, maintain your thoughts in accord with the imagery of the life that you have elected . This is so important that it can not be only too emphasized. The source, God, it works to perfection, giving you what you visualize exactly and you consider that. Exactly. Neither more, neither less. Therefore it is important for you to have imagery and obvious thoughts like the crystal consistently.

For example, a great many people want to have a beautiful car. The error that they commit is to convert in goal to have a beautiful car. ¿What do they hope that the universe give them to? Physically Auto does not have no so-called Beautiful car.

I know clearing like the crystal! You visualize the car exactly, that marks, model, color, options and everything. You see the agency where sell it ó, look for it in the Internet and know ó exactly so exactly I eat be you possible that he is what you wish to have like beautiful car. Visualize it all days at the most think on him, but the fence will be of you. This is an efficient way to imagine. Imagery of the mind expressed are the life. ¿Do you want to travel to another country for some dream vacations? You see the travel agency.

And you get the whole bag of tricks, you get brochures, itineraries of flight, precise, hotels, reservations of rented car. Construct all of the journey, exactly, in your mind in full color and with every detail.

Life is an expression of the internal imagery. All that we do and we experimented an expression comes from our imagery, of our imagination and of our thoughts. Work on your imagery. In order to change your life, you change your imagination and you change your thoughts.

The life that you are living is the life that you have imagined .

Visualize and each aspect of your life, just as you would like that I go imagines . Do it daily at least during an hour. You the imagery are life to your mind, expressed to the extent of that they are believed.

Walt Disney, against all the odds, critical and you caught a four line stanza of failures to his imagination becoming the worldwide king of entertainment. The mouse Mickey was thrown at the lowest point of his
Failures. Nobody may have forecast that he would be able to forge for oneself an empire as from a mouse. When we want to go with will where you carry our more valued imagination to us, we became creative forces that do not have limit. Albert Einstein said that imagination is the creative force more powerful. Learn how to want to go where your imagination and your desires take you.

If you want to increase the power of your visualization, utilize the eight hours of the day that you use to sleep. It is easy to learn how to stay alert in your dreams.

In a status like that one, you can be the author of your of one's own dreams, painting your perfect life and living on during your time fantasy.

Remember the subconscious mind can not tell apart between real experiences ó imagined. Dreams are the best form of visualization. This ability the Shown Off Sueño names himself and has numerous books that they show like doing it.

Have many reasons to live and to be rich. How many more reasons have, more rich you are going to turn around. Instead of having like only reason to make a fortune to be able to pay your accounts, you look for reasons like traveling, collecting objects of art, buying houses, having exquisite clothing, expressing your friends and family to share good things, to buy a ship, to take care of the ambient midway and suchlike things your love to the other ones with gifts, invitations. You visualize these reasons with precision. The reasons convince the subconscious mind, to the heart, to interiorize the goal. In like manner they give you material of work. How many more reasons have, more powerful it will be your imagination and more rapids will be results.
Hedge you in with imagery of things that you like, You Get brochures and magazines from cars, houses, places, things, bodies, sports and activities that you like . At the most the whites of egg be the imagery to your vision, plus clarity and rapid you will catch up with them.

Dream at large, maintain the dream the sufficiently height and consistent and the dream has to be shown at form, by law. For universal law you can not fail. Do not get worried like. Everything is fixed for you while you make out your part of thinking, acting, speaking and being according to that dream.

Simply dream, visualize and next you begin to do something. Make the following thing that you sense that you should make according to the dream and that keeps on forward. A little more that you have not forecast and this follows and that keeps on until it is completed shoots the little that you do .

Are you dedicating him for less 30 minutes of your day concentrated only in visualization and thinking up your future? The imagery are the life to your mind expressed. Imagination is what the life designs .

Dedicate a specific time for imagination and also maintain those good loud visions in your mind all day long.

The world is an illusion, as we will see much more clearly as we advanced . Look at it like an illusion, believe that it is an illusion, a designed illusion to be a tool to know and to experiment To Be.

Once that way you do it, you will remember always from creating for oneself the following version of the illusion that you wish to have. Literally you create the following version of the world painting the imagery your mind with certainty. The world does not have because being as he is ; Simply he is that way because all of us agree to that.

Your subconscious mind can not see the difference between an experience vividly thought up and a real experience.

The visualization creates one new subconscious. You visualize your more crazy dreams and fantasy objects. - live in big! Back him up with faith, belief and certainty.

The subconscious mind is where yourself image is stored.

Command your pansies concientes and your actions. Reprogram it using visualization and belief and you will reprogram your world. Your world is your revealed auto-image so that you may experience it.

Your imagery, your imaginations are your life, expressed in physical things. More precisely, the public demonstration is in the world physical the visions in them what else you believe, good and bad, concientes ó unconscious.

How doing the impossible: Quit than fly your imagination.

Get enthused with your imagery. Energize them with positive emotions.

Emotions are moving energy. Gratitude is an one belonging to the most powerful ones belonging to these emotions. Inculcate each cell of your body with your imagery. Instill each cell of your body with gratitude that you right now have attained your goal, although you not be experiencing it at this moment. You have received it automatically through the perfect laws of the universe (before you ask it for right now you have been given it).

When you act, remember putting this same energy to your actions.

You elect imagery. The Source, God, you do the work effortlessly and in shape perfect. The agreement is that one.

.

The mind concient creates impressions in the subconscious. You do it through the repetition. The subconscious express mind next those imagery which causes an expression in the physical manifestation. The mind conciente does not cause expression. Only you can print on the subconscious using repetition and faith, belief, certainty. The reason which he is this for that way conciente is the mind all classroom of thoughts maintains – those that you create and those you do not believe in. You are like a savage monkey, jumping with thought in thought all day long. The subconscious, in addition, what you regard as truth for you maintains only, that you believe . You maintain auspiciantes to the thoughts. For that reason only the subconscious, or that some call the heart, cause expression.

Your subconscious only take the affirmation and no the refusal. If you visualize putting a goal I no longer am poorer only you will take the scant word. Instead, change to something the same way that I am rich, rich and abundant the no does not interiorize The subconscious, do not interiorize denied sentences. Only you interiorize I am something, not I am not something .

The repetition is to can. Concentration also is to can.

You are visualizing your goals, look at yourself to yourself in possession of what you desire . Say and think I am, not I will be . This is I criticize . "I am " he is scientifically this way. The Ahora is the only moment that exists, all the other music illusions.

Maintaining constants everything them besides variables, an individual ó a society they will catch up with the riches and happiness if they have positive and grand imagery of the correct way.

Now that you right now have covered up the basic concepts of the creation, let's move bigger our things progressively. Get with you what you know on imagery, since you will need it to keep on constructing your understanding of the riches and the bigger lessons will continue to be this that they will allow understanding still but deeply these knowledge in the shape of experiences, in scientific form and spiritually. Let's go on

Thinking and talking : The instructions of and the thoughts are for life In The Neighborhood related to imagery. Imagery of the mind expressed are the life and also the thoughts belong to the mind expressed. In other words, life is the expressed mind. The densest party is your reality outside mind, without no separation.

The separation enter you and your world outside is illusory. As we moved forward in this book, you will be able to see of first coat like this he is that way, scientist and spiritually.

Life is your expressed mind. The world outside is the part but dense of your Ser; An extension is mind to your. When finally you stop believing in the illusion of separation your power will increase dramatically. But enclosure now, your you can still to change your mind to change your world. Now you will see as this happens and as we advance you will see scientific and another- type proofs. Finally the illusion of separation will decrease rapidly totally and you will have attained his mastery. But still at this moment, simply you empower you with knowing it.

This chapter is a course of how thinking of manners that they are friendly to the universe and to his laws, you form of creating as a whole the universe per se in intentional form to obtain one you want to live in and no the one that at present you find . They will mention the laws of the universe several times in this chapter. These are laws that you will learn in this book and no necessarily in this chapter only. The jigsaw puzzle's pieces will stick as you read .

Very well, let's become absorbed in the mind. You the densest part is world to your mind

Your thoughts are the life, expressed. Let's repeat what we learned about imagery. Life, The Source utilizes your thoughts like the instructions with which creating your reality in the world material. The life expresses your thoughts in physical reality. To express is to make somebody know, it is to bear witness, articular, to communicate and to transmit. The force of Life makes your thoughts known to you and to everybody else forming experiences for them and objects that can be felt here in the world physical. You experiment your own thoughts.

The imagery of your mind of such way than may give you tell of as they are adequate and as no. This is the way you know yourself in the way that you experiment is yourself, it to yourself and it is the same way that you grow . This world is designed to afford to experiment To Be. It is designed to afford to experience an idea, his effects and consequences.

Your status of external riches is to your an extension and testament been of interior riches. How obvious and true you are in your thoughts of riches you expose yourself externally.

The life does not select than thoughts expressing and as no expressing .

How would you do it? Therefore, expresses them to all insofar as you have them and you believe in them. You have free free will. This free will is really free due to the fact that everything is executed without filters neither favoritisms. Free will is really free for the fact that in reality you get results all the time, not sometimes, and you do it in exact form.

Inasmuch as your thoughts not be concientes, deliberate concentrates on any theme of life, you will be affected by the outcome of the thoughts of someone else. Inasmuch as your thoughts be clear, concentrates and no contradictory, your results will accelerate . Very few people are able to accomplish what the people call miracles simply with thinking on a very way's thing and strongly. For them.

His intentions be able to no happening like they they desire they do not go by the head to to them not even for single moment.

The suffering stems from the incorrect thought. An indication becomes of being out of harmony with the laws of the universe. The only purpose behind the existence of suffering is to warn a person that its thought is wrong and to alert her of the existence of a thought

More lifted that you benefit him. The suffering stops as soon as you find a road but lifted, that more lofty thought. In the presence of suffering, do not try to resist. Instead of it, examine with an open mind open and the answer will appear to you without fail.

The mass meeting gives more power to the thoughts and bigger velocity to attain goals.

Your dreams, thoughts and visions will forge your world. You will rise and you will step down with the ups and downs of your thoughts.

The repetition generates integration and internalization. You read this material repeatedly in order to attain wisdom and think of the correct way time after time. Through the repetition, the concepts get embedded in your subconscious. They interbreed and they become yourself.

You can predict the future examining your pansies, words and present-day actions and applying the law of cause and effect.
Never the answers are denied to your questions. Any question that you formulate seriously.

Believing, she will be for responded to you exactly. Neither more, neither less. If sincerely and believing questions like earning one million dollars, the universe will conspire to bring you the knowledge, the tools, the people and the events in order to deliver you the answer. If you ask like to earn thousand million dollars, you will also get the answer for the value from that quantity. Einstein was not born being a physical mathematical temper. The, simply formulated the correct questions in shape credible. As you see, the universe works for a perfect law that never fails neither that favors no person in particular. Once you understand the rules of the universe, deeply complex and at the same time simple, perfectly balanced, you can not fail to have success in predictable form. When you see chaos and impredecibilidad in the universe, simply you are seeing something that you do not understand but it is something that is organized and is predictable certain laws.

Nothing is difficult for The Source, God. Perfection and equilibrium are the nature of The Source. So that all laws are applicable equally, universally and without fail. Formulate the correct questions, in credible form and honestly.

I know specific and stop of seeming in your mind. All thoughts count and pay off . Changing one's mind all the time confuses the universe. Imagine that you go to the travel agency saying I Want To travel. After you stare to the employee with the blank mind. The employee is ready to do the reservation but you can not do her until you tell him where you want to travel. Imagine that you say it Good, I would like to go to Moscow and to Timbuktu at the same time.

Again, the employee can fail to satisfy the order. Now imagine that you tell him: "Good, do me the reservation for Moscow. No, no, wait . Do me the reservation for Timbuktu. No, wait, I am not sure if I am going to be able to pay. No, I do not can . Not, perhaps I do not want to go to those places ó such time neither although I want to travel " This is the way many people think in all day long. The universe gets mislaid with those thoughts like in the event of the travel agent and therefore produces confusing results.

Anything that you put him the attention to receives energy and it grows .

Withdraw his attention and you die . I know conciente and deliberate at this moment. The intention goes with attention. What you propose and you begin to become manifest what you give him attention to.

We became what we considered . We are the total indictment to our thoughts.

Everything her laws of nature work always; They never fail, not even once only, job than if that way it happen would cause chaos in the universe. The nature of the Source is the perfection. You become what you consider, without fail. If you think about the riches, without contradictory thoughts, you make a fortune, without fail.

The matter is materialized thought.
Remain always still. A dead calm mind is not affected by situations that cause fear ó anxiety. Never perform on ó in a status of confusion anxiety – the thoughts that you may have in that been music.

Erratic and suicide. Calm down first and remind you yourself to the laws of the universe before proceeding .

The thoughts attract similar thoughts. The souls attract the souls. The minds attract the similar minds. It is a dance for the round trip. The way it works in is that one. Also you are right to say that thoughts attract similar matter and that the matter is thought conferred an objective aspect to, this is, thought once object was turned into. In like manner the body and the world are a denser extension of the mind and the mind an extension but subtle of the body and worldly. They are not separate. Use this knowledge to understand and to create your surroundings, your riches and your health again.

Do not permit that your mind live in dissatisfaction, even if at this moment you feel unsatisfied. If you live in dissatisfaction, the one thing that you achieve is to feed the maintenance of the conditions of dissatisfaction.

You change your attitude, look to those circumstances as you unlace perfect of your thought in the past and an opportunity to return to create you to new.

Thank them for this gift.

The combination of intention, belief and detachment enables to move you calmly for the life, knowing that the universe always fulfills your intentions but you make it out of the best-suited way using a sequence and an intelligence that surpasses our understanding. The life begins to work and you begin to smile, amazed of how simply everything works for you. When you confront a challenge ó a problem.

You can relax because you know that that is a part of the outcome that you proposed and that they go for to you to get solved to you. Enclosure elks of than appear the problem I defy ó right now it was previously solved. Appear to go down a step away more close to the outcome that you proposed . You will find with that focus that life works for you automatically based in desires and intentions that you back up with belief. As you practice this, you earn more confidence and relaxation with this new form to live causing that the ancient spine-chilling habits and anxiety gradually crumble down and begin to have success more and more rapidly. As you advance and you become your Ser's more conciente the time between the thought and the physical manifestation will be reduced more and more.

Even you can get to disappear, there have been and at present there are people that have that capability.

The more incredible envelope life, just as many have discovered it, is that the act of creation is enough similar to the act to realize that something right now was there. The universe is designed that way. All Aquí is all, Now, what may exist. A person is able to experience what right now is there through the election and that election you become more possible for the increase of his understandings and perceptions of life.

Have intentions but no preferences ó addictions. Elect unlace them futures by means of your intentions and desires but you accept all of the present moments. The present moments are the perfect outcome of your last thoughts, statuses and actions. They are a gift to afford to experiment to you.

Resisting and cursing the present only they cause that his nature go on and on .

Talk to you to yourself and formulate your questions to you, expecting an answer.

Soon you will take his hand to obtain answers that come of inside.

They come like sentiments and no words but you can understand what they mean and you perceive the heart of the matter ó the employer of totality theirs.

Learn how to tell apart between genuine sentiments and thoughts ó emotions that are masked like sentiments.

When you set out to do something, be a very mind. Do not have substitutes ó preferences and you practice the detachment at present. What is happening to you in the present since you have brought it with your previous thoughts, words, actions and statuses of being and is here only in order that you see yourself accepts with joy to yourself, like a giant-sized living mirror.

That what you stand against, persist . Do not desire that your present be different .

Live happy in the gift. I know your present mind but do elections on your future and you propose your future with a mind, a very mind with concentration and precision.

Never quit of learning .

Observe and you affirm the truth, and the truth will do you free. If you are bankrupt, say it that way. Look him in the face to the eyes. Admit it. Next you look at the errors seeing than they caused the bankruptcy in your thought.

You always affirm the truth, job than what you resist, persist, you confront that and you bring in the light you liberate you. But I know careful in admitting it. To say Before it is more correct I Had thoughts, actions and statuses of the being that led me to The manifestation of the bankruptcy that now I am observing saying I am bankrupt . In fact, never say, think you seat ó that. The declarations of I am ó I am they are always manifested for the universe.

Change the cause, no the effect. The thought is the cause. What's physical is the effect. Trying to change the effect in direct form does that you hit your head against the wall. For example, if your sales decreased, it is not because your marketing is incorrect. Would be able to be, for example, ó toward an aspect that you have an incorrect attitude toward your business The People have attitudes such like Detest to have That to go to work of the same, I Hate this work, I Hate this work of customer service, If only you may relax me and doing nothing All Day Long. If you have these attitudes and you ask yourself because reason the things do not work for you do what you do in your business, those attitudes would be able to be the cause. You go deep deep, find out and you examine your status and thoughts. Always music causes it.

Here there is something you can require a certain time in order that you sound like you. The understanding of this will give you an enormous power. Here we go : Remember we saw that you are in the point of cause on the things that happen in your world. Those that have to me.

Once the time to study and to experiment was taken this they know it for experience that each person causes 100 % of her world. Now, remember the thought takes time to become manifest in the world physical depending on the matter in point. Therefore, What is that then are you looking when you use the eyes to see? ¿What is what your holes see ? If you are in the point of cause for everything you see that and the thought takes time to become manifest, obviously you are looking at oneself to yourself in several stages of your past. Read this again.

Everything this is an illusion. A designed illusion to allow to see yourself so that you may create version of yourself, in order that you may design your growth next time. What you see with your eyes is an illusion designed from several stages of your past. Your I real, Being, you go much more forward. Your thoughts of this moment are very fence of your I real, running
Behind only a fraction. Your I real it is what the thought causes, of the you per se non-verbal he is of where the pansies, the observer, the soul leave . But another theme is that one. Let's revert to the illusions and I eat to use them to create riches. The way to use this world sighted with your eyes is to see it like you passed and to use it deliberately to observe your Being and to see that it is what you have to change and where growing . The only way to use the illusion is that one, which is a gift that they have given up to you gently in order that you may see oneself to yourself.

The results of the present are effect of previous thoughts. In other words, when you look at the world of today, in reality you are looking at oneself to yourself like eras yesterday.

Reflex the world of this present moment your thoughts and status of the being before the present moment of the now. It is all an illusion. Úsala as such and you will be always rich and marvellously.

Madness is to do the same thing repeatedly and to expect different results. ¿What are you doing? ¿Are you doing the same thing every day and expecting a different result? If so, laugh well, now, with a loud laughter, next you change from now on.

Yes, exists a so-called thing thought auspiciante. It is the thought behind the thought. The thought auspiciante has more creative force than the patronized thought. What normally we called thought it is the patronized thought in reality. You examine your thoughts in detail. They have an auspiciante, a cause. If you think that you need to eat, that thought has a cause behind. That one is the thought auspiciante, in this case, is the belief that you are hungry .

Begin to become aware of this. Ó comes from the subconscious mind that the people call the heart. The thought auspiciante is the way you really believe in and that hold like truth. It is your subconscious programming. If your thought and your thought auspiciante about something they are not coordinated, else they agree, the thought auspiciante wins . For that reason it is that the people that prays imploring God never get hold of that their prayers be answered because his thoughts auspiciantes are of not having ó to lack. Give you account of that and correct your thoughts auspiciantes.

The science has proven that temporarily the party of the brain that you tell you where you finish puts out the deep meditation. Meditate and you will attain the vastness of the mind that never you considered and the conscience like possible. The solutions for your business, the commercial new ideas and strategies of riches are only a fraction of what you will begin to happen. It has been said than if not you go away stops inside you go away stops appraise .

Do not judge, do not condemn . You cause negativeness to you and that maintains that you judge ó convictions. Cause delays and a great quantity of unproductive results.

The expanded riches cause the expanded perception. The pardon causes the expansion of your perception. When you forgive the other ones and to yourself for what you think that it was done incorrectly, you open up to see what you and the other ones really are . You open up to see the beauty and the capability that you failed to notice such time. You become more tolerant and you believe in the freedom and love.

You begin to live less in the influence of the circumstances. Many things happen that they expand your perception. The expanded perception expands your conscience, your abilities, your opportunities, your contacts and a great deal more than what you drive the riches.

You become what you consider and in what you create most of the time.

According to some calculations, you have around 50,000 thoughts per day. Thoughts that they do you are some to walk ó scratching ó to control your physiology.

Other ones are unconscious fantasy objects. Many are repetitive and only concients are a few and deliberate. Begin to observe your thoughts, to be awake to they. Do not permit that they be at random like until now. Do thoughts more and more concientes and intentional.

Becoming conciente of your thoughts, you will wake up and you will become an intentional designer of your life. You become conciente simply making a decision to be conciente.

Turn your mind into a sunny and beautiful island that you allow you to having only thoughts and positive influences and ó puts out any negative thought negative influence. Act like if you have Positive Alerta's Force of Mental Defensa that is on duty in your head the 24 hours fending off internal negativeness and day girl. You have to earn all of the encounters.

Form a group of discussion. He is a group of people that they think equal than you. Meet frequently to exchange ideas, materials and motivation. Where two ó but they are assembled, the power and the whole are multiplied not only that that you add her up of you split them. A group of discussion is very powerful.

Multiply the power out of every participant enormously.

The instant that a negative pansy appears to you, say " Alto!, And immediately move over to a positive thought. Do not feed the negative thought neither for a second. Remember this is unlike the resistance to negative thoughts. What you resist persists . There is a great difference.

The negative influences can come of friends, of the television, of the news, of your imagination, from seeing things, etc. When you see those negative influences and the appearing of his correspondent thoughts you should put them out instantly. There is no implicated resistance. But when you find that negativeness is so loud that you have to resist her, that simply you can not avoid her without a struggle, at that point you should approach her head on. Be unable to bear with her. Instead of it, bring her in the light, look her in the face and examine it with detachment. You try to look into one point by sight separated from in order to see that it is compound, because you appear and because you are affecting you of that form. Crumble it, understand her and find out that he is what propels her. Ask yourself because you are there, that he is what really he is . Observing her conscientiously, going into his composition and finding the root of his causes you will be able to surpass her.

First one comes with the mind. We got on the mountain Everest, first with the mind. We got to the Moon, first with the mind. You learn how to walk, first with the mind. Arrive first with the mind where you want to go. That simple. Whatever you wish to have, have it in the first mind. If you want to have an enormous new house, possess her first in the mind with exactness of details. Live in the house, first in the mind and physics will continue to be to it in shape to him.

As much as it be you possible little Eve to read ó to watch bad news for included television if you think that it is good for your business. The bad news create bad imagery that interfere with your great vision. The world appears to you like you you choose to see it.

Bad news are often a prophecy that tends to come true for their own nature.

Now you know that your thoughts create your reality but you make a mistake if you try to link all of the today's reality in day with the thoughts of today. The present-day thoughts affect some things today but some of the things but " sólidas"en your reality of today they stem from many days of last thoughts. Depending on the theme, the mass meeting and certainty, the thoughts take variable quantities of time to be shown at physical form.

Think on your thoughts. You think about the thinking things. You watch your pansies and I know intentional with them. Thoughts are eternal and forever. You can propose and creating in the past almost as efficiently as you do it normally proposing ó having intentions for your future. People's majority never they consider this possibility in spite of the fact that you are useful and powerful. This capability leans in the eternal nature of the Spirit and in the eternal nature of the quantum parcels that the building blocks of the universe are .
What right now you have become physical can be changed by the thought but that is much more difficult that to turn physical what not yet is physical into something.

What does he mean praying without pause? This considers It has been said than still before you ask it for you have been given it. It has also been said ask for and it will be given to you. ¿Do you perceive it? To ask for is not to implore, you do not beg the Source, since you right now have conceded it to you before you ask it for. Imploring and wanting simply brings the scarcity to you of what you beg and you want .

This is not a spiritual idea simply; One can try at the quantum field. The quantum soup contains all of the possibilities of everything literally. Now, Here. In other words, the spiritual promise is also a scientific fact.

Let's go on . It has been said that also you can make anything if you create it. It is said in like manner than anything you pay attention to him to and that intention manifests itself to. The way that you create things as from the quantum soup of pure energy in is that one – by means of attention, the intention and belief.

Many experts, teachers and learned persons of all the cultures and epoches have told us not to get worried, that we practice the detachment, that we rely on the mysterious functioning of the universe you did not sole because the universe runs on supreme intelligence but also because perhaps you not give yourself account of that your soul or being superior is electing .

Here the significance to pray incessantly matches then: You have a desire; You want you to become reality, that you become manifest . You propose it with the whole heart, clearly, attentively and with certainty. You pass this intention to the way's Source that you know . The simple intention is sufficient, but meditation and quietness are more effective when you have the time to practice them.

Then you pass your intention and you peel back yourself of her. In other words, you look at her from behind the scene without the particular desire of than happen of a specific way ó in a specific sequence of time.

Because you know that this works, because of your certainty and consistency, detachment and gratitude, you will become manifest of most unexpected and miraculous the way. That is prayer.

The incessant prayer happens all day long, every day, with such intention focused for all of the desires of your life, with certainty, detachment and gratitude. It is not something that you do once every once in a while in a red-letter day a determined hour where after you go away and you act totally different and once the rest of the day was mistaken for. It is something that becomes a lifestyle.

That is the incessant prayer. Prayer is the midway to be active, ever-present and it departs from the normal state of vigil. You are creative in tandem with the Source; Take to the action, you have character. It is not periodic, passive, unprotected and with the emotional attraction many of us were with educated from children. Ask for and it will be given to you, look for and you will find, before you ask it for it will be for granted to you. Asserting car, designer as a whole, ever-present, thanked, made to receive and not give imploring – the incessant prayer is that one.

Forget about the head that is the same thing as prayer beg him to God that he do you a favor and for the one that God decides if ó concedes it to you not, after reduce and you do so much your work like the part that he is in for . Prayer is will, will in reality co creative – and the part that you are in for to you is to have the obvious intention, the certainty, the gratitude and the detachment. You have to have certainty, you have to know that you have this power and you must trust that the Source is friendly with your plans. Inasmuch as that way you do it, your sentences, your will will be answered.

Opinion does not exist at all for the one that The Source decides ó no conceding the prayers. The laws of the universe apply to wholes equally and without fail. Prayer is an energetic internal process, a call with a detached expectation of answer without a pinch of doubt. It is a loud and certain will. When you realize that the order not even is necessary, that you are one with all that Is, with all whatever you wish to have, with the grantor of that desire and the carrier of that desire, you will be praying in reality and receiving without fail. Granted that your prayers will be of pure gratitude for what right now you have been given before you ask it for, the request is not necessary . Simply I know thanked and smile .

Play right along, have fun with Life! Life is joy. The essence of life is Joy. Where there is joy, the creation is abundant. It is easier to create riches where there is joy and joy gives the riches significance.

You consider what in the mind and you feel in the heart you are in what you convert yourself.

All that you see to your around the idea belonged to somebody. Look around yours. Nothing can exist before existing first like the idea of any mind.

The man becomes what you consider . – The Bible

Your mind is infinite.

Leaving behind constants everything them besides variables, an individual ó they will become society rich and happy insofar as they have positive and grand thoughts.

Well, we right now have covered another building block. You know like to consider and talking about the correct way for the creation of the riches. We have to cover up a step more at the thought's field. We saw as the imagery work . Imagery are an aspect of the thought. Next we continued to cover everything it relative to the thought per se. Now we will see the latter aspect of the thought: Let's go on

Goals:
The map of route toward and in the riches.

There is a determined way that you should put your goals in. ¿In order that does it suppose that they are goals? The goals put you in the on the way to the correct thought toward the riches. The goals concentrate your thoughts, they give them format of the correct way for the universe and maintain your imagery in consistent form and no at random. So that there is a correct way to get goals.

A lot of envelope has said the importance to put goals itself. This is not nothing new. Regardless of this, what you are going to see now is something in all probability new for you. You will see as you should put goals to catch up with the riches rapidly. It is sufficiently good to get goals but to put them correctly you are tremendously powerful.

Welcome to the world of the powerful goals! Your goals are the prophecies of that in what one day you will convert yourself.

Else you know where you go, you will finish exactly there. where? If you fail to plan, you are planning failing . Else you have specific goals, you will not obtain specific results. Remember, The Source takes from your pansies, the imagery your mind and manifests them in reality.

You can see as this works exactly, scientifically, in Carnal Cuántica's chapter. But your thoughts and imagery are the working drawings for which your world is designed and created for and for you.

Goals are planned thoughts, guided thoughts. Without thought planned and directed, your life will be unplanned and lost appearing like risky and reliable no.

Simply you have to know where you go and the answers of how arriving you will be the host for with his own time. Don't worry. Simply believe that you will arrive .

Your ideas, visions and dreams, anyone that go, are the prophecies of what one day you will catch up with and in what you will convert yourself. You can change your morning changing your inner being today.

Take seriously goals. A study of twenty years of duration that was implicating students of an one belonging to the universities of a group known as Ivy League, famous for his academic prestige, the lives of all a graduates's litter tracked . 3 % of that classroom had his goals written at the beginning of the study.

After 20 years, that 3 % that began with written goals and plans had a value in money but lifted than everybody else understood in the 97 remaining %. In like manner, that 3 % declared itself to have more satisfaction and joy in its lives.

Praise seers and the dreamers are saviors and the driving force behind the world. Inventors, artists, philosophers, educators, learned persons, businessmen, designers, scientists, leaders, etc are these. – Whichever you dream big and believe . The world lives in his ideas and can not quit that his ideas die off without bearing fruits. The world is beautiful due to the dreamers and seers. The world and the universe completely back up these dreams with the only condition that the dreamer creates it completely and act in consequence. The universe, the world, The Source, God is friendly of great help for your dreams and aspirations. Simply have a vision and you will without fail catch up with her inasmuch as you believe that he will that way be . Everything else is to your favor. ¡Dream at large then! ¡In reality, dream very, very at large!

Conformity can be your worse enemy. If you do what the multitude does, you will obtain what the multitude obtains . Every day the people get up in the morning, he is going to be worked up like everybody and you do things in the same way in that everybody else do it. Ask them because they do it and simply they do not know how to respond. Simply they work hard like does it everybody and his reasons stops to work they are that they wish to pay accounts and to supply support m same and for his families.

If you have 30 years, you work hard like the rest, like the rest of the multitude and that is all that you do, easily you can say, in brought near form, where you are going to be in my whole life when you have 50 years simply with seeing the people than today in day has 50 years. ¿Maybe does most people of 50 elderly years have financial independence? No. They are not the great majority neither rich neither financially independent. But they would be able to be it.

Only with having stopped to come next after the multitude because the way it is done in is that one.

You have to make a very thing in a different way in order to get something what the multitude obtains better from: This is, having specific goals, reasons, visions and conscience of the riches. Reading this book, you have begun something different and cause of riches. Practicing it daily, you complete the requests of the riches.

Talking in realist form, one can be rich and financially independent still to the 18 years even younger ó. Everything depends of how soon, so strong, so precise and believing one begin to get goals, to have a vision and to construct a conscience of the riches. The key to catch up with the riches is not only to work hard. The people that has made a fortune are with and without working hard. The key is the conscience of the riches as a whole and the map is to get goals of the correct way.

An error that the people commit once it attains goals exists . This error sinks them back. The best form to understand this error is illustrating it.

Let's say that a person has the goal to have soon one million dollars in the bench's account and at this moment they have four thousand dollars. Then they do everything correct, goals go down, they visualize, they get certainty, they act and they do all good. They maintain tall the vision of the million dollars and they manage to have it. Once you tell her sign that one million dollars have the bench, they get happy, of course and there you begin the error. The error is that they begin to look at the bench's account trying from maintaining the precious and ancient goal, getting worried when you get to a level underneath the level of the goal and suchlike things. They change the sight toward a not attained goal and they begin to look toward the last goal that they attained . They begin to live in the past, trying from maintaining the past.

After they begin to get worried and this keeps on and keeps on until they lose the million dollars. They stop doing what they were doing in the beginning when they wanted to attain the goal and they begin to live as they used to live before.

Always put your goals above than your present reality. This does not mean that you never are going to be pleased with your achievements. This does not mean that you have to become converted in a miser. No. On the contrary, it means that you should enjoy every day without worrying you. You should not worry about losing what you got . You should not worry about maintaining the million dollars that recently you have gotten . Instead of it, the million, wins have a good time earning it and having it but when you achieve it do not transpose you to worry about keeping on. Put a higher goal and make obvious in that new goal enjoying the success without no worry to maintain it.

In any event, getting worried is a foolish thing to say right now than the worry same it is what you sink you.

The conscience of the riches and all his activities are a lifestyle, no something that is done occasionally. Your thoughts, being and goals always should be in the biggest version of yourself, not in the last version.

Have care when putting goals and visions for you. The heart, otherwise called subconscious, the tendency has to drink and to interiorize the part not denied of a vision ó of a declaration. For example, if one of your goals is never again to stop to pay your accounts in date, to formulate the goal like Nunca further to you can turn out to be pernicious I am going to stop To Pay accounts because only you can be taken to stop to pay accounts because the fear that that declaration produces upon saying it is evoked. Instead of it, modify her in order that he be always there is quite enough money to live the life that I elect, in order to have a good time and for to invest

If you fail to plan you are planning failing . Plan, establish goals and visualize them. You dominate this ability.

Maintain your goals always beyond your zone of comfort. If you attain all the goals and you fail to create bigger goals, you stop growing .

That can be dangerous in spite of the fact that you be comfortable . ¿Did you know that they begin the people's majority to have symptoms of diseases correlated with the old age within the first months to go into retirement?

They indicate the brain and the body that life is about to finish, that the society no longer requires over its services and that certain shows right now can begin to go off through retirement. The problem is not the retirement ó withdrawal but the signs that the people accept that that one is the problem.

Reiterating, there is no bad nothing with retirement but be careful with the signs. The lack of goals can be pernicious for your health unless what you want be that. The goals do not have because being over money ó the profession. There are countless valuable things to get goals all over the world and they have nothing to do with ó making money a race, from personal goals such like sports, journeys, hobbies to goals of worldwide reach such like the protection of the ambient midway and the beneficial organizations.

You never need to know how you are going to attain the goal. Do your part, liberate him and you will achieve it.

Do not live neither by accident neither default. Live for design. You design your life utilizing goals, visualization, imagination and plans, all fact in consistent, quotidian form, clearly, with precision and in detail.

It is not sufficient with getting goals. Goals also must be established of a certain form, a form that the friendliest be possible with the laws of the universe. The correct steps to get goals are these:

1) You write a list of what he would like to have to you, doing and being from now on and during the next 30 years. List all that you may come up with, child and big.

Places to visit, things to have, residences, experiences, companions, abilities ó skillful actions to be acquired, things to do, people to know, projects, charity works, health, habits - Everything! This is not a list of what you think that you can achieve. A list comes from what you would give you her but incredible life to your pleasure, whether you think than can achieve it ó not.

A life that is fantastic incredibly for you. Your list should have for less 100 things – it is not difficult to join for less 100 things in 30 years. In order to be very rich, you have to have a list of around 5,000 things – still the smallest details on your desires have to be in that list.

2) For each goal that you included in the list, you write the reasons which you want for to have them. If, for example, you wish to have a large house, you write the reasons for which you want to have it. In other words, What is what you are going to do with the house and within doors? Dramatize . The reasons empower your goal and do her easier to imagine, to visualize and to be enough. He feels life and they do that the subconscious welcome it much better.

3) You get clippings of magazines, brochures, of Internet, photos, etc.
Of things in your goals and stick them on your daily newspaper. A daily newspaper of your visualizations begins in paper form ó in the computer. Glue imagery of the things that you want to have – cars, actions of the bag, buildings, properties, journeys, clothing ó anything. Look at her followed – it is two up-to-date times commendable.

At the most real and detailed be your visualization and imagination, all the more quickly and you will attain your goals with more precision. Imagery are very important to have in my whole life.

4) Every day you read the list, you look at the photos and go by less 20 minutes, two times per day imagining, encouraging, visualizing all your goals in detail. If you meditate, which is highly commendable, practical the visualization of your goals in meditation. The meditation puts the more possible fence of The Source you can be about to plant the seed of your visions at the field of the infinite possibilities of the creation and in, the best place to you.

Next, Here, Now do something that you bring you closer to your goal. Always there is something to do now although you be little . You will open the next step toward you with you, a step than puede being invisible until one accomplishes the first action. Each act is an act of auto-definition and of creation. Act deliberately and conscientiously of such way than each act bring you near to your goals, no longer there. Act with purpose.

Do everything with gratitude. Think, talk and act with gratitude, the gratitude to know that you have the guaranteed success if you act in accordance with laws. Gratitude is a declaration of certainty. That is to can. I know genuinely thanked and enthusiastic for the fact that you right now have achieved your goals, since the warrantees for an universal law are and he will that way be. This type of gratitude brings about miracles.

7) ¡You enjoy fruits, you enjoy experiencing your goals when they be shown at your reality! It is safe that it will that way be guaranteed by an universal law.

Make sure that the daily newspaper of your goals and visualizations be easily portable.

There is no use having a daily newspaper that only remains in house. You should register also in him, important thoughts in addition to experiences and conditions. Register as much as ó when you turn out to be more convenient to you be possible for you at the end of the day. A daily newspaper helps very much to discover you and to create you with precision. Do not worry about on the structure the daily newspaper – simply make it out of way that you be easy to use yourself.

You write also the problems that you find in your daily newspaper. Suffering is an error in the thought as you will be able to see in the chapter Conditions. Writing your suffering helps to evaluate it clearly and to discover the error. You write your sentiments also – it is important because sentiments are communications of the soul, the closer part to The Source. Do not mix up sentiments with emotions and thoughts – I know careful with this.

When you have good ideas and inspirations, write them off.

Perhaps the best hour to obtain inspiration is pregnant of vigil when you wake up in the morning.

Instead of taking immediately your baggage and to plan the day ó whatever it is, continue to be lying-down and question thinking on something big that you would like to know and the answer will come to you clearly. Tightly when you wake up, before you be wide awake, relax and softly formulate the more important questions of your life to you, softly without waking you up. You will receive the answers of incredible manners, in forms that they would not do it during the rest of the day. Aside from this, when you begin to practice what is in this book, they will begin to receive great quantity of ideas of all classroom of sources and apparently by coincidence. Have your daily newspaper at hand. Do not wait for writing later; Write now before you forget ó get lost to you. Do not forget to register the dreams that you have to the night - you will one day find utility and everything in the correct time for them. Dreams are not only imagery that you receive when you are asleep. The people think about the status of vigil like the status to be alive and when they sleep like the state of rest. They think that all decisions and useful activities are done when they are awake. Well then, what's amazing is than you, your Ser, your soul, spirit ó like you want to call him, never sleep . Your spirit, soul ó any other term which you be comfortable with, that never ever you sleep with. Simply change conscientious, dimensions statuses of conscience.

You are multidimensional. The status of vigil is a status ó dimension, the dream it is another status and there are much mores. They all count ; All have influence your life in the status of vigil and vice versa. Although you do not create it, remember you tell all of the thoughts, that dreams are thoughts and therefore they tell and they influence the events of your life.

In reality, you never sleep, in a manner of speaking. You are a soul with a body and no a body with a soul. Insofar as you are conciente of this and you are awake to this, you will have over everything.

You can do the following in order to make easy the future consultation to your daily newspaper:
1. Having parts separated for different kinds of information.
2. Making an index at the end.
3. Using many-colored lapiceras.
4. Giving the pages that you want to mark tangs.
5. Or, inventing another system that be desirable to you and that easy access guarantee you to the information any.

Make a revision and your daily newspaper rereads at least once a month ó preferably more often. Once a year, you read your daily newspapers. What happens when you read is than, all at once, you begin to see your life from a new perspective. You see successes that never you thought that you had had, you see things that you should change and errors see that it is necessary to correct. All objective to maintain a daily newspaper is to be able to see it in retrospect and to find amazing opportunities to understand yourself better, to see more clearly and to grow faster.

When you write on the daily newspaper, note down the exact date, the hour and the place.

That helps you to detect employers, tendencies and velocities.

As much as the principal be the quantity of events and experiences that you grab in your daily newspaper, the principal will be the benefit that you get of the same.

It converts in habit to maintain the daily newspaper.

When you write and speak of your goals, write on present tense, I Am I am I am that Does Not Have Importance be trying from creating from now on in your life, you find a way to lay eggs I am ó I am . For example, if you wish to lose weight, do not say I am going to Slim 5 kilos I want ó To Reduce 5 kilos. Instead, you can say, thinking and writing Estoy 5 more thin kg. Do the same thing with the riches. The only moment that exists in the universe, so much scientist as spiritually it is the Ahora. From there you come then give it the I am .

Do not get worried be more than enough how they are going to obey your goals. There are powerful forces that operate in the nature with an intelligence and an infinite coordination. The things, the people, the books, the places, the television programs, the films, etc will begin to appear and will help you to attain your goals. In other words, coincidences will appear to you. Simply you visualize your goals in credible form.

Put out all the other thoughts that they deny your goals.

The trick is in the details and in consistency. For example, if the goal that you have is to have a new house, write it in detail.

Write where the house is located, how many you have rooms, the size of the lot, the size of the house, the furniture, etc. From now on visualize it that way and not change to seem – this is very important . Understand that the universe is in motion manifesting all your thoughts in physical form.

Each one their you become an a certain amount of physical form somehow and somewhere. You change to look alike, you undo the work. Simply maintain the thought until you become totally physical.

The best date as a mere formality it is the goals in the Ahora's eternal moment. Enclosure if you are putting a goal that you would like that it be obeyed from here to 10 years, declare it and consider it in the present tense. Putting an arbitrary date in the future introduces a status to want and to wait. The wait impedes the achievement. The arbitrary dates also interfere with the natural velocity of the universe usually we do not know which. When you say for intentions of the coming year I am going to be worth millions, how do you know that you would be able to not have convert in millionaire to endings of the month that you come ? Then, the only time and the real place that is in the universe is Here, Now.

I know exact. Define your goals and visions in exact form.

Else you aim at nothing, keep calm because you will not with certainty obtain specific nothing. Talent, intelligence and hard work without exact goals lead often to frustration.

It takes time to manifest them to the thoughts in the physical reality.

They consider the people's majority only upon a few months forward. Today they can consider I Need To buy a ready house and after they begin to reinforce it and to fight to get it. That one is the focus to the short-term chance. Try with the long term.

Imagine the power to put goals 30 years in advance. Listing and visualizing all that always would have you liked daily to have, you set in motion forces of the power of the thought of the universe and well in advance. Still if not you wish to buy a house at this moment, in the meantime know than a day would buy it you would be able to desire ó to buy it, you begin to visualize it now. It takes time to become manifest to the thoughts in physical reality therefore, as soon as possible begin, so much the better. Life after it begins to work automatically.

You begin to obey your goals effortlessly and without the pressure of the time. The things simply fall in their place. Remember than in your goals of 30 years, in spite of the fact that you think that you need a certain amount of here 25 years, establish and you think about that goal in the present tense, Now. "I am " and not I will have . The universe will be occupied of the appositeness for you.

You should know toward where you want to go if you wish to make sure that you are going to arrive. Otherwise, you are not going to arrive. Have goals, plans, imagery and visions. Maintain daily newspapers to register them, to track them and to refine them. Aimless, plans, imagery and visions you do not go to ó to go a long way you are not going to arrive it far away that you love one another with your capability.
How many goals should you have? You never can have over enough. The richest people have hundred of goals. Some have thousands of goals. So many goals have some that you would take 200 their ó 300 years in order that everything become manifest to them. You should indicate to have at the very least five thousand goals. You understand the nature of the goals;

(1) Goals are imagery of the mind that the universe uses to create, the substance. Therefore how many more goals have, more material you give the universe to work and you swim the difficult ó for The Source is impossible.
(2) The Goals have the tendency to become reality in the moments and more unexpected sequences. How many more goals have, your experience of life will be more rich.
(3) When you attain a goal, his proxy disappears . You do not have right now a motive force and the universe right now not you have with working . Then how many more goals have, so much the better.

A person with a goal will achieve less than one with 100 goals. A person with 100 goals will achieve less than one with 1000 goals. As much as minus goals you have, you will achieve minus results. At the most you lay eggs, you obtain more.

How it may be that do you have 5000 goals? Easy. You think about the details of everything. List everything, still the smaller things like having lilies at my garden, decorating the house to my grandmother, piloting in a supersonic plane, passing the night at the hotel Ritz, doing a donation to the wild life's bottom of conservation, buying a convertible BMW, a jeep, a jet plane, a ship, having an aquarium of water put salt in in the living of my house, giving as a present my father an equipment of golf clubs, buying books for the orphan children, traveling to the Great Wall Of China, knowing this people, coming out with this person's type, working with this people's type, suchlike shoes, this designer's clothing, this type of chairs of this I negotiate, these ski boots, " these places in these countries, etc. ¡You never are going to remain without things that you would like to do, to be and to see in this planet!

At the most have them in your conscience, all the more you are going to come across coincidences in your life that they do that they become reality. The life begins to work magically and you have passion and enthusiasm. Remember, the riches are expressed abundance. Again, the riches are abundance expressed, once what's physical was translated, manifested. All that is is abundance, it is your true nature and the nature of Life. When you think about riches, do not think only upon money and business. You consider in everything, all that you would wish to have, doing, being ó to see in your life and in the life of another beings, all life. I have the bigger element of understanding here: You are an observer simply. The life yields life by herself. In any event, everything exists .

All that you are beams to observe and experimenting . Try to understand this. You are an observer with a body to experience your observations. It is what you elect to observe what you determine that what you observe is and you experiment . Everything is there for you, automatically. In consequence, you elect many things and you will see many things.

Something else that you should understand on the goals and thoughts it is the following: Your goals result from you and for the other ones and vice versa. In other words, when you mean to have a ship you provoke an inspiration in a prepared person to construct ships and causing that that person get involved in the business of the ships. This in like manner inspires all of the necessary events and the middlemen that are required to bring the event to have a ship to you. They all win . ¿What is what you think that the fact that this book be written for you caused ?

The effect is of your previous desires to be richer, the desire of thousand of million someone else to be rich and my desire to spread out riches and rich being. He is owed to you I desire that a thing is created and all that you desire comes into being – you do not create anything for oneself without your desire -. You can say correctly that you wrote this book.

His life opens into life; You are an observer with a body to experience your observations. The way it works in is that one. "You observe what you wish to observe " You Experiment than and I eat to elect your experience. How many more goals have, you will allow more to the life that it work across yours for the good of all. The maxim goal of life is to express itself to herself and does it effortlessly, following your intentions exactly and beliefs. Once you see this in clear form, you will know, beyond the shadow of a doubt that, what you wish to have wishes to be with you more than what you wish to be with that. Her you do not desire something else that to express oneself to herself.

Logically, do not be timid, have many goals!

Do not commit the error to put your goals only in terms of money. The riches are the expressed abundance. Abundance means plenitude of everything. Money is only an aspect of the riches. A great many people fail to make a fortune because they plan his goals only upon terms of money. They say things like earning enough money to buy a car ó to earn one million dollars to buy a house. There is a very large error in that type of goals.

A similar error is to have enormous financial goals and very few goals of another type. Some people can put the goal to earn thousand million dollars and to have some themselves few goals more. ¿Why are both focuses wrong? Try to understand this clearly: Imagery of the mind are the life, expressed. He is that simple and Life also is that precise. You obtain the necessary amount of money to manifest the more precise imagery of your mind automatically. In spite of the fact that you may think that money is the one and only form of obtaining a thing that would be pleasing to you to have, you know the life that there are many other manners to achieve it, aside from buying at cash. Therefore, if you have very few goals and imagery of another type that they not be give them money, it is not going to there be a lot of material in order that life operate

In order to illustrate this, imagine that it be possible to open the mind of a person to examine it. Now imagine that two fictional people, Bjorg and Li and the fact that both wanted to become rich be . Li wanted to have thousand million dollars and that is in the one thing that they thought. When examining them, Li's mind also showed few imagery of something else. Enclosure things related to his business I work ó, there were very few imagery ó goals related to the quantity of customers, quality, products and suchlike things.

Also there were very few imagery and goals in all the other aspects of his life. Li had only a loud desire and goal to become worth millions.

Borg also wanted to be worth millions but unlike Li, he had developed also many interests and desires in all the aspects of his life.

His mind, to examine it, was full of vibrating imagery of one a great variety of something else. Even you had specific petty details such like the clothing that you were wanting to have, places to travel, the decoration of the office, manners in that you wanted them his customers, gifts that you were wishing to give his loved ones as a present and the rest of the world, etc to treat. Then you ask her he is : If we leave behind constants all other variables Who looks to you like he became richer, of an easier way and apparently by coincidence and luck?

Bjorg, of course, will be richer and he will be to it of an easier way.

The life guarantees that to you all imagery of the mind that you support should come true like true and obvious. It is OK to wish to have thousand million dollars, but exactly How do we visualize thousand million dollars?! As you can see, there is not enough her of visualization of a lifestyle it is the cause a great many people fail for when wanting to obey his financial goals. To visualize it is very difficult and maintaining an image of thousand million dollars but it is extremely easy to visualize a multimillionaire's lifestyle! Do not separate your financial life and your goals from the rest of the life, since the financial goals are only the midway toward an end, no the end as such same.

Remember, money is a shade of the value, a medium of exchange. Your goals should be in the value of the interchange, no in the money that the shade is .

I know the state of mind of The Source, The Life, God and only say to oneself yes to yourself, never say not. Then the only questions are:

What is what you are asking for? ¿Do you believe in that? Still that way, never ask for, never have the intention because before you ask it for right now you have been given it. Whatever you ask for, whatever you propose believing, she will be yours .

Sightless we perish . – The Bible

Join of the bigger reasons which the people remain for without riches it is that his goals decrease and the imagery of the mind fall down . Sometimes this happens when something new come to your lives and they forget about the original passion that did them rich in the beginning. This new thing would be able to be a little boy's birth, finding the companion of romance, achieving the comfort specially when a person is native of an environment of scarcity and it becomes delicious, etc. The mailbags are no one of these things, but to know it is good and maintaining at clearing what if occasionally you take a seat going I lower ravine you examine your goals and mental imagery. This is a powerful beginning to find out that it is what you spend with your life since imagery of the mind are life, expressed.
You read many magazines of all the themes. The magazines give ideas, imagery, goals, desires to you and much more. How many you have more imagery of your mind, your life is going to be more rich.

If we leave behind constants everything them besides we can say things than an individual ó the riches and happiness inasmuch as they have big and positive goals of the correct way will achieve society.
This concludes our journey toward the thinking of the correct way stops to achieve the riches.

The imagery, thoughts and goals have to be of a certain way. After they have to be inserted within an ampler perspective. In reality, thinking is the second step of the process of creation but he is the one that is in more often implicated actively the riches. However, never ever forget you the step is two of the creation of riches that to think. You are very important that you always remember that. And then is Which One the first step? The first step is To Be, The First Cause. That is what the thought causes .

He is of there of where it gushes forth the thought. The thought would not exist without Being . Let's sink still more deep

Being : First Causa, the beginning

To be is a status, like being happy. You can not explain a status, neither you can do a status. Only you can be a status. You can not do happiness; Only you can be happiness ó being happy. The creation works of the following way: Being cause the thought, the thought causes doing, which lays eggs to the system to receive and to experiment you have created that in being and in thinking. To be is the First Cause.

To be is what the thought causes . Then when you are happy, you have happy thoughts. The thoughts spring up of being. In other words, the thoughts are patronized to be. In fact, nothing happens without being .

As you will see in later chapters, even the circumstances happen due to the status of being and not the other way around like you consider most things.

The happy circumstances do not gladden you. Being happy cause happy circumstances.

The unfortunate circumstances only appear to try your status of preexisting unhappiness. For the moment in that you finish read the chapters on Cause and Effect and on Influencia of the circumstances, you will see clear like this he is that way.

The first step stops to experiment an enormous riches are to be rich. Being rich is an interior status. Nothing has nothing to do with the world outside. The interior status of riches is a decision that you take at this moment and in the one that you become, at this moment. You need not nothing of yourself to take that decision. Once you make the decision to be rich, you make a fortune . It is difficult to talk about this since only you can be a status (you can not do a status neither speaking a status). Now we will try to talk about this status and when you finish this chapter, you will see clear like works being and as you can be rich in this moment.

I know rich, Now, Here, that easy
I know the riches. Do not try to make a fortune. I know rich. You consider that to the happiness in order to help to you to understanding. Do not try to obtain happiness, I Know happy.
Can you see it? You can, or else trying to do things that they gladden you, ó you can make a decision instantly to be happy, simply make a decision . You right now have done this before sometime of your life. Has everybody sometime said " You Know What? I am not going to be given that this bother me.

I am going to be happy and I am going to stop worrying for this theme " What is easier?

Trying to do things ó being something instantly and being given that the status to be allow doing things that be consistent with the status? Of course, it is easier to be happy that to try from being happy. The same thing happens with the riches. I know rich. You will follow him automatically, while you always be, all the times, To Be the riches to depart from everything there. Simply I Know rich.

As you will be able to see through this book and in Carnal Cuántica's chapter, all that may exist exists in the Ahora's eternal moment, Here. Even right now the version of yourself that is delicious and the riches are experiencing exists – if you are not living the riches at this moment, simply are not conciente alert ó to her. When you choose to be in a certain status at this moment, here, now, active the more fast forces of the creation. What happens from now on is that immediately you outmode your conscience to the other one I that in this case the rich self is . To be is the fastest way to create things since you cause an instantaneous change. The material manifestation appears instantly inasmuch as you become the status without doubt. Reiterating, inasmuch as you had become that status without doubt, your reality you will move around rapidly to reflect it. As impossibility that may sound you now, will have sense when you consider the real functioning of this universe, specially after reading Carnal Cuántica's chapters and Influence of the circumstances.

The fastest the way to create is simply to be, at this moment, whatever you want to create.

Do not deny it with the thought. Do not think about that. Simply I know . To depart from all your pansies there, words and actions would belong in mail with the new status that you elected being then. Else you are rich and you want to become rich, simply you make a decision at this moment to be rich from now on. Do not begin to think on it, simply you choose to be. To depart of there, all your thoughts, words and actions should be give them a rich person. In the maximum reality, this is not a lie. Anyway you are everything in spite of the fact that you may be experiencing children in reality sections altogether. Electing to be something further, causes that your surroundings and circumstances change to experience the new status.

Remember, the thoughts come from being. A status to be is what the thought causes . A status of hunger causes thoughts of hunger. To be is the First Cause. Being is, the thought is to do. Being simply is . Being does not take time to be while you take time to the thought to turn the things into being. Then, the road faster you are to be rich, now, instantly toward the riches. Your being changes rich one. Do it that way, making a decision, at this moment and knowing that you are that way with certainty, against any proof, unlike your physical world (which is an illusion anyway). Do it with certainty and clarity and keep on maintaining that state to be. You know that you are rich because in the maximum reality you are very rich really. You will rapidly begin to this riches experiment, hastily in truth. Not many achieve it because they doubt, but we all have the capacity to do it.

Simply bear witness, with full knowledge and certainty, I am and immediately afterwards do not begin to consider it again because you will introduce delay and doubt . This is what I am saying : ¿Do you remember of last time that you were in bankruptcy very sad ó? Did not think conciente on that status, asking oneself if he is real ó not, if it is that you were in bankruptcy ó not, if he is than were sad ó not. Simply you quit for established that you were as you were and you believed him without interrogations. Simply you were that status and that was everything.

Simply you were that. Now, then, do it with the riches. I know the riches, believe in her and turn her into that he is that way, without questions. The universe will obey .

The sequence of creation is as follows from now on: The field united the fact that no is physical (God, The Source) classes to oneself in units physicses not of conscience (individualized spirits ó nourishing of several beings and objects). These units take physical individualized forms next – objects, the people and the beings that see .
Whatever it may be you see physically you come into being in this sequence to depart of, of and for Everything Source that It Is, God. You are creative co in this sequence of creation – you create in tandem with The Source, in the same sequence.
With regard to the riches, therefore, you can check than so useless you prove to be to look at (like your mind, thoughts and status of being) That to your individualized physical appearances you are to look at the result and to quit that you influence the cause (I eat to be your bank account, your material riches, your body, etc and being given that what you see influence your aspects not physical individualized.

You are to amplify to the one thing that you carry you the physical status. For example, if you look around and you see that you are bankrupt and you quit that you dictate that on you that you are a bankrupt being and you keep on thinking from one point of reference of bankruptcy and you have thoughts of bankruptcy and thoughts of incapacity, because you will continue to be bankrupt.

Correct the way to do this is never to face the floor, never look at physique neither accents to it than dictate you that that is what you are . You are not your circumstances simply you cause circumstances. The right thing is what if you are bankrupt simply choose to the more big vision of yourself, a vision of riches and maintain that status to be rich, those constant and ever-present thoughts of riches. Act like if you be rich, what what's physical evidence, without having importance does not have importance than so bankrupt your physical world appears . This will give turn what's physical to match it up with your thoughts and status to be. Always remember, what's physical goes after the spiritual and mental. The design comes from the universe.
A form to create things in this world is by means of the thoughts, the words and the action. But this is the slow road. The fastest the way is changing status. Being . For example, when you say that you have hunger that is a status. When you have trust, that is a status. You are . You need not nothing of you to be a certain status. In order to acquire riches, you are very much, much more efficient to be in a status of riches, turning your being into one with the riches, feeling rich and thinking in rich. Talking and acting in rich. You are going to prove to be difficult rich being if you feel poor and next being thought, talking and acting I eat rich.

Your status, as you take a seat about yourself, your being, your declarations of I am – the way is that one but fast to acquire riches. The way in changing your status is to decide to change the status. Simple. You can do it at this moment. You are like not being happy and simply deciding that you are tired of no to be happy and simply deciding to be happy. Everybody has done this occasionally. Now do it with the riches.

You hope to experience an enormous success! Always the status of knowledge that you have maintains and you experience abundance. The expectations like these, the knowledge to this level, cause attraction and withdraw the repulsion. This is extremely important. You hope to have an enormous success.
Recognize to be powerful.

You can remember the past and looking toward the future, but only you can be yourself here and now. Your being can be only Here, Now. Being and manifestation stand alone Here And Now. Million people consume their trapped status of vigil in their awake dreams, worries and another thoughts of the moment do not present . They are awake but they do not become aware of completely his surroundings Here, Now. ¡Wake up ! ¡You smell the coffee! This simple awakening will produce an amazing change in your life.
Attempt it, commit yourself to waking up, one day at the same time. This combined with I am in present tense, thought and visualization is an amazing amplifier to achieve your desires.
Your world is what you are . Everything in the universe makes up its mind within your interior experience.

Little it has importance as he is out, since a reflection is everything to your own status of conscience. All that has importance is what you are inside yours, since all outside will be reflected and colored according to that. – Have a meal Allen

In order to be rich outside, I know rich on the inside using this type of information in books like this. In order to attain multimillionaire's status, lift your imagery and certainty to equivalent levels and act with the correspondent certainty and being your purpose the fact that you inculcate your actions.
The world is within you. Swim neither nobody reins in your ó you accelerate you except yourself. Insofar as you recognize this, you will change your world. The riches are much more easy from acquiring so that you imagine yourself . It is simple. The most difficult part of the acquisition of riches is to tame your own mind, which is totally within your own control.

It has been said that correctly success is something that you attract for the kind of person you become converted in. What is to be? You can not be planning to be, you can not perform on being, you can not speak being, only you can experience being . Being the quality to be, simply he is . To be is conscience. Being does not belong to a mind. In fact sometimes I mentioned her it can destroy to be what you wish to be. To be is something than your you choose to be, at this moment Now. No longer delay but Now. When you begin to think about it you destroy it. Once you are, you are . Any thought should not after that be on if you are ó you are not - it should be conducive to fulfill that status to be, to live it instead of determining it. He is very often a good idea no having mind. I know still.

The thought has its importance. The thought is a tool like is it an arm ó a leg. You do not use legs all the time. You use her when it is necessary . The mind is a powerful tool.

You do possible a great deals things but she is so powerful that sometimes you take our control. The mind should be used only when it is necessary and 10 % of the time is it. The investigation evidences that about 90 % of our thoughts are repetitive . Worries on the future or going back to live the past are the majority. This is clearly unnecessary. The only real moment is the Ahora. To escape the Ahora is the cause of so much tension, I fail and problems in our world. Your normal status through the day should be of I did not mention . You should be an observer, no a thinker.

You should watch your mind. Just as you observe the things out of you, you begin to observe also your thoughts. This way, you stop getting under the control of the mind. You stop identifying with the mind and it is identified your Being, the being that everything knows it. You begin to live in the gift instead of the past ó instead of the imaginary future. Your tensions disappear and your success blooms . However, do not judge your mind neither curse her if you realize that you have been under your control. A wonderful tool if we learn is like using her correctly. You right now know like using her. Use her only to formulate your intentions, in order to give the appropriate imagery to bring new experiences to your moment of today to life (no within five minutes but Now this very minute). You will begin to realize that than you do not have a problem upon the Ahora, you have events but no problems.

The problems exist in the mind, in your thoughts. How events, music only events that happen and change as soon as they happen . All your problems, if it is that there is any one, are imaginary and are in the future. If still you are alive reading this book, you will know that you never could escape from suffering through the Ahora's moment. If you are here, reading this, you never failed an now is that moment, you never stopped passing successfully through the now.

Enclosure death, the maximum fear for many people, you are not a problem . Those that know what death is really know also that you are not a problem and as a consequence they do not fear him. Nothing in the now is a problem . Nothing Now is a problem – you are designed to play the Ahora in perfect form, but the moment that you begin to worry about the future, identifying with the mind instead of using her, you begin to have problems. Remember, the future does not exist . You are in your mind. Enclosure when you think on the future, you think about him, Now. When in reality you get to him, you do it in the now, no then. For the moment in that in reality you be in your future, it will be the Ahora.

I know observant about your thoughts. You are not your mind. Your mind is a powerful and beautiful tool, but do not identify with her. Use her for thinking of the correct way and put it out if not be to her using . In truth, most of the time you do not need to use her. You think about the following:
Were you be occasionally in a situation at the risk of death that you showed up unexpectedly? ¿What did he pass at that point? Your mind be something but largely it was off .

To be, Being, the commando drank and you managed the situation of the best and more intelligent way. In a real emergency, when the mind does not have spent a lot of time without thinking about nothing, you are present at the Ahora and never it is a problem when you are present . In fact, you become extremely still. Now here the good news come : You are not in need of an emergency to agree to that calm supra intelligence. You can learn how to be there, all the time, to be present all the time. That is true Being. To be is I did not mention . To be is the quality to be, the presence, finding out, the conscience, the Ahora.

Your thoughts, words and actions reflect your to render an account to you, your conscience, your being. You change your being and you will change the world. You can change your being of two forms. A form, you are simply to elect Now to be what you wish to be and to maintain it. You form other, you are to act like if you be what you chose to be. Act as such, talk as such, think as such and afternoon premature ó your being will yield turn itself and it will do the same thing.

Words I am music powerful. Be careful to that you tie them. What you are claiming has means to search to you and to complain to you. – A.L. Kitselman
Quitting you fix them besides things, an individual and a society will catch up with the riches and happiness in the measure than in that they have big and positive statuses of the being of the correct way.
What you have read on being thinks . There is enough to assimilate but he is easy to do, he is simple. In the meantime you want to do it, like a little boy you want to learn new things of the world, to believe,

You are going to find that it is easy to understand to be and to change your being. Inasmuch as you want to be like a little boy, to simplify and to get things done because it is so, you will be able to be able to understand and to interiorize this chapter.

Do not worry about on significance some of these teachings that you have just reading . Some of his significances, the real significances, they will get clarified after you had read all of the book. They are explicable using another concepts. The chapters on Influence of the circumstances, Being, one, Causa and Efecto, Tiempo and Carnal Cuántica will deepen your understanding to be and how it works literally. For the moment, nevertheless, let's go deep into the following stage of creation.

To be is the first step, the First Cause. After you follow the thinking, after talking and writing words and finally acting . Talk in the meantime you think . It is not going to there be a chapter on talking because words are expressed thoughts simply. Simply you watch what you have said to yourself on thinking and apply to it to talk. But because we do not have a chapter separated on words it does not mean for them not to be important. Words are very important – they are expressed thoughts and they have dreadful impact on the creation.

We will proceed to performing on the correct the way to welcome the Mr.s of the riches. Let's deepen more within the world of the riches

Acting :
What receives .

Once I had a dream . I was sitting in 15 cm's little bench of height when an old man, friendly but the vigorous man gave lessons established in a higher bench to me and saying me: "You can not move the world with actions. Only with the word ". And there you ran out of the dream. It took me like two years to understand what you had wanted to tell me. Now I understand it sufficiently well in order that it work . Through experience, proofs and a lot of reading finally got clarified .

This comes from what is treated: Actions are the latter component of the process of creation. It is a little like the swimming of competition. You can not earn Olympicses if only you know like to push water vigorously and almost going from a tip to the other one. You are acting and very strongly. Nobody is going to disqualify for lack of effort. You would for sure be able to obtain the highest grade in effort, but the people that earns gold medals in the Olympicses of swimming are the one that gets ready in the spirit. The being is prepared. They have trust, motivation, concentration.

His minds also are ready, his technicians are ready . They are rested and alert. The action at the swimming pool is the most visible portion of winner's total parcel but the final part becomes of creating that winning race of the gold medal.

In your life, in spite of the fact that you be able not to know it, you create your first experiences in your I, spirit, being, next in your mind, next in your words and finally for your actions. Begin in being and next you move over to the thought, next toward the word and next toward the action.

In fact, the action only puts the necessary system to receive in its place and experiencing what you believe in being, thought and word. They do not concentrate the people's majority and they nourish putting the intention in his first it is three steps to be, to think and to speak the one thing that they do to work like madmen all day long and after they ask cause they do not have success . They do not use the word. The Word refers to the laws of the universe it works in, the way the universe, in all the levels, not only the physical and visible universe. Those laws of the universe are not a God's order. They are the laws that regulate and enable the functioning of the universe simply. Those are the laws that they work not only spiritually, but also they have been proven scientifically by means of the Carnal Cuántica. The Word, those laws, they have nothing to do with no religion, person or that fuere in individual the laws of the universe apply to anyone equally and without fail and to everything all the time. They never fail . Those are laws I eat such give them cause and effect (call karma ó sowing and harvesting in the spiritual teachings ó the laws of conservation of energy in science) and music once several parts of this book were dealt in.

The actions are a part of The Word, The Laws of the Universe, but they are only a component child. It is important to know as the role that they play is actions and I eat to use them for the creation of riches and of anything else. Actions are important, for sure, but you have to understand than the latter dried music. His role is to receive what right now you have created in the three previous levels. First you create and after you receive your creations, next you live them.

The action is for receiving and to experiment. You create a business in your I, your being, next in your thoughts and words and finally you perform on putting a system into operation to receive this business of a physical manifested way of form such that you may experience it. ¿Can you see it? The action does not create . Only you receive and experiment .

In fact, when it is tried to experience, again, the action you do not create the experience – the mind does it. The action only helps the mind to make the thing, next the mind makes a decision I am going to live this experience positively, negatively, with joy, with fear, rapidly, slowly ...

Know this and you will stop pushing water at the swimming pool

Always do a job excellent. Always concentrate on the task and produce an excellent value of her. Do always this no matter that the task seem so little. Still the smaller action has the potential to be the cause of the bigger opportunity that comes toward you. In this universe, the cause for something else is everything and it is in turn caused for something more. Still your smaller actions can be the cause of something big that it has not used to be revealed and that you need .
Still a so small action like a smile and a good turn can cause the beginning of a relation that a door turn on itself that never you considered and like possible.
The ideas do not have value without the action.
The action does possible to receive what comes in your road as from your intentions.

The intention initiates the creation of riches, the action enables its reception. Take action .

Still the smallest action can be what you need to move yourself toward a great riches. Everything tells, each act counts, each act your following world defines . The universe is an enormous chain reaction.

Do not try to do nothing. Simply do it ó, do not do it. Never try of doing it. Or you begin to make ó for it you do not do it but never put you to try to do it. If you try to the universe do something you are going to try to deliver you a result. But if simply you do something with the determination that he is going to be done (not as the fact that it can get to work rather is going to work), the universe you will obey that determination and you will return to you with his own determination.

You right now know like to do the things that you are not doing . If you be earning $100.000 a year you know that you know like to take that to $ 1,000,000. If you sit down to be thought you would realize that you have some plans, some clues that can take ó to that at least to put yourself in the road to that level of more lofty entrances. In other words, you can not say really Not himself the fact that it is the first thing that should make my entrances to increase Is Not Possible . You have a clue of at least the first step, than, for child that he be, he is all that you need to begin with and the rest will make itself known as you advance . But else you do it, else you are acting on the first clue you do not go to arrive at a walking pace two . Cover the breach now, doing what you know, now. Begin now. Simply do it.

Else you are in the perfect business, do not wait for estarlo. Begin now in the one that you are in and move your he would like to have you to gradually. This is all the same with the position, knowledge, etc. Begin now; Do not wait until things be perfect before beginning to act.

You stop reacting and you begin to create.

I eat if. Act like if you be the person that you want to be. Act like if it be impossible not to arrive where you want to go.

The opportunities increase as they take advantage – Sun Tzu

Derive profit from the opportunity that is further at close hand and will open the way previously occult toward but opportunities. For law of cause and effect, making good use of the closer opportunity, you will cause the display of a lot more opportunities previously balanced out for you.

Quitting them besides constant things, an individual ó society, they will achieve riches and happiness inasmuch as they undertake big actions of the correct way.

As you can see, it is not difficult to act. In fact, the easiest the way is of the creation. In the past, you have exaggerated it, but as you can see, step of a very large system is the latter. This only, being ignorant of the previous step when acting, explain very much about because many people are not the rich and happy that they would desire . Now you know it.

But always remember acting is very important in the chain of events, although the leader not come from the chain. Not be ignorant of it now. The road to the riches and to happiness is the perfect equilibrium's road. Balance your body, mind and soul. Balance the time and the accent that you give being, to thinking, when talking and you did not sole when acting act all day long at the same time that you maintain a thought and a visualization little structured, for example. Neither you should go by all the time in spiritual nutrition and being ignorant of acting and everything else. That not only would be selfish rather you would fail to complete the cycle of creation.

Now that you have the complete set of the tools of creation, let's examine the fuel that does them to work. The tools of creation are only that: Tools. They need a last ingredient to work. A so powerful ingredient that no other force can compare it

Certainty:
The more powerful force and antidote against the failure.

Certainty, faith, belief is a necessary part to create riches ó anything else. It is what you give him to the universe the clear track to make what you wish to do. As you see, you can not become a status without having certainty, right now than without her that one becoming converted would be not to become converted. You can not be happy if you are uncertain of that you are happy . You can not create goals without certainty, no only of his precision but also that reality is going to turn around to you.

Inclusive talking and acting doubtfully does not have to might.

Many teachers of many religions in all the ages have taught that we have faith, that we have certainty. This is not new. But now you will check cause always they have taught us this and seeing like to create and expanding your faith, something that until now it has been shy for many people.

Remember, as you read, that faith is very similar to a status, to be. You can not be spoken to of faith, testifying, only you can be faithful, safe. The way to do it is simply to make a decision to be safe, that simple, no quitting that another contradictions come to you. Anyway, let's proceed and this will become easier and more obvious.

The last part to get faith is to understand as the unIverse works .

In Carnal Cuántica's, Tiempo's and Causa's and Efecto's chapters, you will see like works the universe and this you will give you the faith since now you know what happens behind the scene exactly. Once you understand as it works, you will believe .

Everything is possible insofar as you have certainty

Inasmuch as you have faith and clarity of
Thought, things are possible impossible ó, but in reality, nothing is impossible.

Believe . In reality, I know sure.

Persistence, feed the faith. You can use the persistence to increase your faith and through faith you have persistence.

Persisting, still when it seems like you should be beaten, your faith increases in the outcome and attracts it. This is a decision conciente that you take since faith enables the persistence. It is a tight circle. You can not achieve too much if you persist but you continue telling to oneself that it is not going to work. Persistence is a step lightly in front of faith since you can use it to construct faith, but each it so happened that persistence takes you have to be continued by a step in faith.

The persistence, literally, yields benefits. Nothing is really impossible.

Eliminate all of the thoughts of doubt and fear. Never feed them not even for an instant. I know diligent, attentive and give account of your thoughts simply deciding it. Every time that you catch yourself doubting ó being afraid, stops short those thoughts, not leave them progressing . Do not foment them but be unable to bear with them. In his place, look at them conciente, detached, like a not compromised observer. You look at what they are, from where they come, because reason come to you and as much as they last to you. Observing of this their way, you will be able to go down behind them, to find out his causes, his dark origins, to put them in the light and finally to eliminate them.

Fear and doubt are the only one enemies of your dreams and of your visions.

Certainty. Still face on in front of proof against, I know sure, believe, have faith.
What if are able to to it? Everything is possible for that one that you believe . – Jesucristo

Confidence. Certainty. Create it totally, in unquestionable form. In God's world, certainty is the one thing that realizes . The substance is of the miracles. This force moves mountains.

You doubt, confusion, fear and worry have partially his roots in a person that what you wish to be and to have does not know .

Opportunities and abilities that you have at this moment, here, are enormous and incalculable. In other words, you can not run out. Your belief is your only real limit.

They will hit you on so much you create . In the real sense, it is not that God reward those that they have faith. The thing is that the universe moves building blocks, the quantum particles, depending on the information that you receive and of the certainty of that information. This is so scientific like spiritual.

Fear is false proof moving for real. In reality, absolutely nothing to be afraid is not, why your Being is had by everything and it is indestructible. Your Ser is designed not to lack nothing since right now everything has it. He is in like manner indestructible, but his manifestation here below, you come with many illusions and one of your purposes is to surpass those illusions. One of those illusions is give it that the abundance does not exist . Nowadays we know in scientific form, thanks to the Carnal Cuántica and to spirituality, that the one thing that is is abundance just as the advisees went for the teachers during centuries.

In the event that you catch yourself fearing something, know that it is an illusion and look for to find out that what is that illusion since in reality there is not is nothing to be afraid.

It is temporary not to have money. Take I get big lessons and opportunities for positive change. Do not be afraid of him not having money.
It is not necessary to get to the point of not having money, but if it is found that you do not have it, don't worry. Look toward the lessons and opportunities that you contain . The fear to run out of money is a terrible disease. You steal the opportunities of growth, impede that the people attempt new things and you maintain them busy. The fear attracts also that to what is feared to him and fear makes the poverty poverty. Still that way nothing to be afraid is not except to fear per se.

When you need to seize yourself more to something, normally it is the best moment to let him go.

Have faith. Believe . Know with certainty. When you take a glass from water to drink, you know about without the minor pinch doubt that you are not going to fail in raiseeing it and drinking the water. Not even they go by the head that you would be able not to be able to drink the water to you. You do it with certainty. The level of faith, belief and certainty that you must have in yourself, in the laws of the universe and in the capabilities of The Source to work perfectly all the time is that one. It is the certainty that you should have on receiving still before you ask it for and in the guaranty that you right now have it. If you think that you do not have something, make a decision now, that you right now have it and you will have it.

If you think that you do not have something, make a decision now, that you right now have it and you will have it. Do not say but I do not have it. Do not deny . In the course of time he will become you natural. Until then, make the best that you can and never think that you do not have it. Your mind keeps watch . You can acquire faith with practice but it is faster to make a decision simply and once and for all that you have it. ¿How to do it? Simply decide it.

How much faith, belief and certainty should you have? You should have that one that is level with knowing . Simply take her like true, just as you know that today you woke up ó that you drank the mentioned before glass of water. In that level, you have so at clearing that something is truth and it will happen follow-up being sure included if physical world appears to something in opponent in him.

This is the triad of the like having whatever you like it : Ask for and you always will be for given to you, look for and you always will find, it kicks back and you will open up to you. But the triad has a catalyst that must be ever-present in order that this work:

Belief, because belief makes everything possible. There are no empty promises. There are no rewards only for the good people. It is the way the universe works in without fail neither exception. The triad and the catalyst can express oneself of the following way: Desire and have the intention and you always will have it. You look for the truth and knowledge and you always will know what you attempt to know. There are no real limits for your growth since you are more than welcome to experience any election that you do .

Still that way you have to have certainty that those declarations are true, because if no be it ó if they be partially truth ó selectively truth, that he is what you will obtain exactly.

When you be of two minds you perform on doing it credible. Else you have trust, act like if you have it and at the end you will turn you that into somebody with trust. Do this with all in that, do not have trust but you wish to have it and that way the circle of your confidence comprising aspects of your life more and more will begin to open.

The persistence, literally, yields results . In like manner you authorize you and reinforce your belief. Persist . Persist but in your persistence, quit that things happen . Quit that life resolve . Don't worry. I know unglued.

Not to know what one wants clearly is an one belonging to the bigger causes of doubt and disbelief.

Faith is the external elixir that gives life, power and action to the impulse of the thought! ¡Faith is the point of start of all the accumulation of riches! ¡The faith underlies all of the miracles and of all of the mysteries that can not be examined for the scientific rules! ¡Faith is the only antidote known against the failure! – Napoleón Hill

You can think and selling the faith to you per se. At this moment, the people that gets worried continuously are be more than enough they consider that and they sell the doubt. In order to create faith, repeat, repeat and you repeat positive affirmations, repeatedly, every day, all day long and the subconscious sometime is going to believe.

Fear is a false proof of something that seems real. He always is that way.

It is not your state of nature. The status of The Source and of your Ser is intrepid, since there is nothing to for me to be able to threaten the Being and swim of that Being lack .

When you be afraid look him in the eye and find your false proof always proof is something false.

Never get worried . Worry is fear, false proof appearing like real, putting imagery in your mind. Imagery in your mind are the life, expressed. The worry and fear put negative imagery in your mind and they create that on which you get worried and to what you fear, an illusion that they convert to you in real and if you go on and on finally you become physics.

Worry is the absence of the necessary facts in a situation in particular. It is in like manner the absence or suspension of certainty.

It is energy wasted and incorrectly canalized.

The best form to remedy the fear and worry is to confront them and to examine them completely, to crumble them in his components and to see where the false proof is . Take consciousness and continue lifting your level to become aware of more courteous details. This increases your confidence while you reveal truths causing that the fears eliminate themselves.

Notice really that it is what works and that he is that out of order and you live the truth doing that for observation works . You observe the truth, know the truth, the truth considers, you speak the truth, you live the truth.

That expedites results and holds off fears.

You know now about on laws the universe that never they miss and you will know more about them in later chapters. You know that these laws guarantee exact results when to to them obvious, detailed, consistent imagery stock up and with the intentionality backed up by certainty and the action.

You know than these laws run on an infinite intelligence that you can not predetermine. You know that also never you should look at physique to it and permitting it that you determine your thoughts since the ones that create what's physical are thoughts. Granted that you know everything Why then would you have to worry yourself this? You examine these declarations again, slowly, bit by bit. You will see that no reason to get worried does not exist absolutely.

Remember than before a problem happen right now it has been solved . Before you ask it for it will be for granted to you. All that may exist right now exists, Now, including all your potential problems and his solutions. All that you do is to move your conscience to experience portions of the same as you do elections. The bigger lessons and opportunities for your evolution come in the worst moments (because suffering is an indication of mistaken thought)

What you need to do is to learn in those moments, everything, to examine where the thought is incorrect and once you do corrections the fruits they are fantastic . Then, why would you have to get worried? ¡There is a reason for it! Never the universe is mistaken. The chaos is very in our mind; The property does not come from the universe. Again, if the universe works with certain laws and those laws never fail and you can know those laws and to apply them, then Why would you have to get worried occasionally? After all, you can predict an outcome based in the application of those laws. The worry only attracts that on which you get worried . Worry is a prophecy that tends to obey itself for its own nature.

In reality I tell you, if somebody tell this mountain, Ve and throw at Your Sea without hesitation in your heart but believing that what you say will happen, it will that way be for granted to him – Jesucristo -

You can do whatever you believe . Whatever you wish to believe that you have, you will have it. In other words, you always have that in what really you create . You think about it. You always have that in what really you create and insofar as you create it. Never the rule is broken off.

Leaving behind constants everything them besides things, an individual ó society they will catch up with the riches and happiness insofar as they have big and positive certainties of the correct way. Now you have the tools of creation and energizante that gives birth to them forces it. Remain always in the status of certainty, and you refuse to being in any other contrary status to the certainty neither have no thought that go against faith.

It is now the moment to move over to more big operations of the universe which you are going to use tools and certainty within, the fields and laws. The first theirs is the law of cause and effect. This law is a beautiful promise that it enables to have guaranteed effects and finding out the causes of everything in the world. You always wanted to know cause the things happen and as you can do that they happen . The first step to know is the study of the law of cause and effect, next seeing the truth on the circumstances and the influence of the circumstances. Let's begin to examine that he is what you have below the hood of what you surround you, you experience that

Cause and effect: The principal law of the universe.

I have the more important law of the universe here. I have a principal key toward the conscience of the riches here. If you live according to this you can not fail to make a fortune. Understanding and living on the law cause and effect guarantees that you will not fail to cause the events that you wish to experience, than predecirás results and you will find out the causes of your situations. Simply you learn this law well and you read the chapter on Circumstances and on success and you will be well in your road toward the riches and the prosperity. The riches is created when you plug in the tools of creation propelled by the faith in the law of cause and effect correctly.

The law of cause and effect is the principal law that it does to work the universe. It is the number one law. All scientific and spiritual teachers have tried to teach it. They may have said that you harvest what you sow, ó that you obtain what you give ó what goes returns.

That each action has his equal reaction and opposed ó many similar concepts. The Carnal Cuántica teaches us as this works, exactly, level subatomic as we will see in subsequent chapters.

I have what we are discovering here: ¡It is multiplicative! In other words, not only you will experience what you cause to others that they experience or his equivalent one day, rather you will do it in multiplicative form!

If you cause that other ones experience riches and happiness, you will turn you to you and you will also experience it but you will experiment very much like reward more of what you caused to the other ones. Of which the life is treated is of the growth. This is that way for any imaginable experience. At some point in the complexity of the continuous space-time in your Vida's some point,, by law, you will experience a multiple of what you cause that other ones experience . Nothing escapes to this law. Even if not you can see, with your once five senses were limited, where it is going on at this moment, you know that it is going on . Use her to create an immense riches

The scientists coincide now that nothing can be observed without being affected by the observer. In fact, they have drawn as a conclusion that still his experiments must come true with a procedure which the identity of the administrators of the experiment hides in and give it the subjects of the same in order to approach precision since the scientists' expectations affect the outcome of the same. Still this can not create an independent experiment.

The thing that is being observed is created and entertain for the observer. The scientific proof, specially in the quantum physics, show that you are in everything point of cause that you see in the world.

Cause that other ones enjoy an immense riches, increase your consciences of the riches and will experiment an immense riches. It faces the life today. Any business that relate with the increase of the people's productivity and connectivity converts itself always in a big business than himself autosustenta. The programs of computers, the nets, the transportation, the electronics and related business improve the productivity, the standard of living and they grow in return. But this is only a scratch on the surface. A magnificence still we will be the host for a lot of principal when we begin to mount business that intentionally are designed to hit instead of receiving . The designed business to cause primarily than them besides people turn around rich. The business of the future will be designed for causing growth that is true growth and no growth with collateral effects in some other place. These business will cause an increase of the standard of living of the population and an elevation in their consciences and well-being. At the most cause that the other ones have riches, you will have so much more riches, effortlessly.

Whatever you wish to get, give her as a present first. It is the road faster. For whatever you wish to have, cause that another one have it first.

You harvest what you sow, Karma. Law of Cause and Effect. This Law never fails and of what you sow ends always harvests.

It is in your better interest to sow good seeds. Do not be unjust with nobody unless you desire that injustice drive you you someday. The hatred, the greed, the avarice, the anger, the negative thoughts and the negative actions provoke the necessary circumstances of suffering to cause a correction in the negative thought. Remember always than all universal system

I complete it is one, One, in spite of the fact that you appear separated. You will see that this is truth as you read this book. What you do to the other ones is what literally you finish becoming to yourself.

Knowing to can of cause and effect that you have the thought, you can predict the future examining the pansy that you have now with precision. Fortunately, we have power to change thoughts and to alter the future.

Cause and effect. Karma. You harvest what you sow . They have given us this message from several sources over the years and this is truth. It works without fail and to ignore this law is the cause of a great deal of our suffering and poverty. He is that simple. Simply I know intentional and diligent envelope you are that, beams, you think or you say . Know that a cause that has an effect is everything. Next ask yourself " Which is the probable effect of this that I am, I think, I mean ó do I make? You ask the answer to this he is what you will harvest later on. Therefore, if you cause suffering to others, you will go back to you in your life's some point. Ignoring this law has caused a lot of suffering in the human race and following this law you have caused also a lot of prosperity. Suffering does not have external mysterious forces that they cause you to at random.

Nothing like bad luck does not exist . Everything within you is individually and within them I he is the family's buses, the company, the community, the country and the world which you inhabit in. Each status to be, thought, word and action was caused for something previously and will cause something after it. It is to come to realize this done and asking that what that thought caused that I had recently is ó as what you will allow you will be the effect of that thought that I am having to do fine tuning in your being and getting in line with the universe. It is the road to experience riches and abundance.

The things get better when you get better . The things aggravate when you aggravate . The world everything is in yours. You are in everything point of cause that happens in your life, be than be ó not conciente of it.

If you are in everything point of cause that happens in your world, and also are to it everybody it means that groups of people collectively are in his collective world's point of cause. Success and failure, the events at the neighborhood, to the wars and the natural disasters happen not owed wholes a very person in the group but because all the people collectively are causing that that collectively is affecting them. This brings our business to us.

You will obtain more fast results if the people you work with and that you correlate yourself music with people with conscience of riches. Remember also a very powerful way to get something is causing that someone else have it.

Combines both and you will see than so largely beneficial you are to insure you than your employees, commercial associates, relatives and inclusive the community and the world in case of being interested have access to the educational material that you allow them to constructing the conscience of the riches within themselves.

Imagine that you are in an island only with a person, working side by side during a year. Imagine that both are very friendly and that they share books and the food that relatives send them among themselves. ¿What would happen if did you slap person to him? That person, the smack would or else return you, would find another way to harm you, would stop sharing books and the food with you ó well would have a friendlier heart toward you although another guy not take from reprisals. Still if the other person not get revenge, he would have a tension accumulated that you would cut short the freedom and comradeship that they were having . This simple experiment shows you that it is impossible to damage somebody unless you return to you of any or another way. You wound somebody and you will eventually cause an injury to yourself. In other words, if you can notice that this is that way Why would you have to want to damage somebody? Still more exact, Why would you have to want to injure yourself right now than when you damage somebody do you damage yourself?

Now then, if you are at work loudly in your business, Why would you have to want to undo the work harming other ones in ó through your business, right now be your customers, employees, suppliers, the society ó the ambient midway? Sticking at what soon we will see on the collective conscience and hers results in the subsequent sections.

Why would you have to be passive when the other ones harm other ones in business? You know that sometime you will affect you in negative form when you permit that a corporation and your Government do business in oppressive and harmful form, then because you would quit that that happen in the meantime do you take a seat in the peripherals waiting for that you receive the negative part of the profits that concern you? Inasmuch as you permit that elections and the conscience of the other ones have bigger importance than them yours own, you will share his consequences. In the measure in than yourself determine and do your own elections on anything, you will enjoy your own consequences. The riches follow those rules, just as they do it them besides things. "I do not care " do not exclude you.

Ask the people that he lived second worldwide war. If the people had not told from the beginning Me I am not interested in, my problem is not, Hitler you had not come that far as you did it and the people would not have to have suffered bombardments and the fall of the economy caused by the war. Stops to be able to exist Hitler needed the collective conscience of I Don't Care, I am separated. If really you are interested in to turn you rich and to keep on being rich you are better than you begin to put care on the world. You are for your own good.

The commerce that you do not forge for yourself based in mutual benefit leads to equilibrium, absence of peace and finally to the war. Be than your be a person, a corporation ó a nation, if you are obtaining super you go away to profits at the expense of the person you are with doing business finish your damaging.

Any way, someday, the same way that you have harmed your commercial member. This is evident in the world and predictable for the law of cause and effect.

Peace is the bigger incentive for prosperity and it is in your own interest, for your own prosperity, to promote the peace. A form to do this is to trade with justice in your own business and provided that you can, to search to correct unjust business in the world. In grade that you have peace, you will have prosperity. In other words, anyone be the level of prosperity in which you be nowadays, that you will be more and more prosperous if you have more peace.

This is truth, so ridiculous as you may sound you, even for those that they are in the business of the arms.

For example, the industry and the military machinery in the world it is one of the older consumers of budgets. Each second and minute spend million dollars in military supplies themselves. Everything this would be unnecessary if there be peace. The military expenses do not go round for economy as the rest of the expenses do it the greater part of. It is slothful money largely. You observe the amazingly expensive nuclear warfares that are getting destroyed now. Those million worn-out dollars may have used up each theirtheir minute in another activities that are really productive and that go round in economy. Still if they wear down to guarantee the survival and equal employment opportunity for the poor persons and disadvantageous about the world, giving them the opportunity to get going a business instead of working all day long for a bread slice, the entire world would be prosperous .

You would be many times more prosperous. Imagine if the thousands of million short people poverty convert him in productive citizens with caning of purchase: ¿Would your business not benefit? This is possible if the military expense be addressed to, in his place, this cause. Those that benefit by now the business of the arms would be in another business in a world that would be dozens of times more prosperous than the present-day.

Peace is prosperity while the war is it not. If you want a living proof you look at United States. Still United States got to fight against oneself until you joined up at a nation and you stopped the wars between his of one's own statuses. Now an approximately equalitarian system of legal protection is prosperous largely due to its internal peace, cooperation and free low commerce for all within its frontiers. The European Union is becoming aware of that and searching to do the same thing. In kind another initiatives around the world, Asia, Latin America, Middle East and Africa all over the world, but in different measure and velocity. Therefore in your of one's own business, the peace through the fair commerce promotes, still when you have opportunity to be unjust. In your community's events, country and world, you see what you may do to promote the fair commerce and peace. This is for your own prosperity.

Nothing in your world is real . Swim of what you see you are real . It is an illusion created for your benefit of such form that you may live your thoughts and statuses of the being of first coat and to refine them, to improve them.

(for that it is said that suffering is an indicator of mistaken thought). Being,
Thoughts, words and actions create it. Your mind changes and you will change the world. You change your being and you will change the world.

Here a shortcut goes in order to solve problems, including financial problems. When you come across a confusing, not very clear situation and problems, look at her and I gave That I am and accept it by heart because you have caused it and separation is an illusion. Next, ask yourself to yourself, Why am I that? All confusion and fear will disappear and I mentioned the automatic solutions they will begin to appear in front of giving oneself account of the I am that. This devotes itself to any situation and in reality not only to a problem.

Have you seen like functional the life like a great and complicated mirror? What you make him the other ones, you do it to yourself. If you want to be happy, gladden the other ones. If you want to be free, liberate the other ones. Use this same beginning in your business and you will be able to elect what you return .

The interconnections between all the life and the system of cause and effect are extremely complex, efficient, effective and transforming. A small cause apparently would be able to have an enormous effect in the future (ó in the past, ó in the present) The physiques have a beautiful way to explain this in simple form. They call his Effect butterfly.

Have a meal Gleick, in his book Caos, como"la explains the Effect Mariposa notion of than a butterfly, whipping the air today in Peking you can turn a system of storm New York next month into " and this is a simple example. A cause is everything and it has an effect. Everything has an effect that had its cause. A giant-sized chain reaction is not tied for the time, space and form. In the event of the riches, the effect is, ó well poverty, ó well riches. Give you account of the thoughts and of his probable effects. A thought in particular it's probable that do ó lead to the riches to poverty? Things do not exist such like a thought in vain, word, action been ó being.

You not feel guilty . Forgive yourself for the failures and you choose to act rightly next time. Evita blames her, since one comes from the concentration's big destroyers and confidence. Do not live in the past. Forgive also the other ones. You do not do them a favor if not you forgive them, neither you do yourself to you neither. They still deem their debt of cause and effect as all their actions (karma), be than your forgive them or no.

But when you forgive them, you get free of a cycle of negative karma and you free your energy toward positive something else also.

What you find in the effect right now is in the cause. Try to understand this intimately and next you live deliberately.

There are interesting something else on the cause and effect once the riches was related here.

From where do you suppose that the people get the inspiration to initiate a business, race, passion or hobby in individual? Definitely, individually they are in the point of cause for his desires but also there is something more. The collective conscience of this planet and universe has also his paper in causing this inspiration. Remember, the life always manifests imagery of the mind and grants each sincere and certain desire without fail any. Mind's field is one (your mind and all minds fix the field united of the mind).

Logically, the reason which you have an idea for is in part because other ones have caused it. This is the way in than the offer and the request work behind the scene. If one million people desire and they believe all of a sudden that they can have a certain accessory of fashion, for example, a person with the level adapted of desire and belief in getting involved in the business of fashion will have an inspiration to create the accessory and to sell it to him to that people. From there then that all desires and beliefs will be satisfied.

The reason which you have an idea for is in part because other ones have caused it.

When you visualize something that you wish to have, somebody is visualizing the same thing like something that they want to ó hit to sell.

Then, next time that you have inspiration, rejoice knowing than a group of people, child big ó, actively you are asking for and hoping that you fulfill his desires.

In other words, somewhere about the world, the people are praying very strongly to have that in that you are inspired . They are the answer for your desires.

Everybody is an answer we all are a prayer Mr.s and miracles for the other one, in spite of the fact that we may see it in immediate form.

Nothing to get worried is not except of the worry same. You have success - that is the one thing that always you have .

A form easy to understand that everybody is a gift and that we all are responsible like a mind united at the mental field you are to think about Hitler. ¿How did Hitler be able to have been a gift for us?

First, realize that he never may have come to the power without the worldly approval as a whole. The world created necessary conditions stops than the go up . When he rose to power, the world said it is not our problem, we do not care about what Hitler be doing with that people while we be all right . What Hitler caused was that separatist ideology, combined with the way that our collective status of being had created a fertile status in in the world in order that he grow and prosper . The you may not have done it only, a very child person against the world. You needed of cooperation conciente and worldly irresponsible person. You can not accuse Hitler from victimizing the world without blaming the world like creator of circumstances to be victimized. Hitler allowed experiencing a negative aspect of ourselves.

Now we are less prone to go into worldwide wars. We know that it is not a good idea. Also we are less prone to be ignorant of the urgent situations of the other ones and to act of separatist and stopped worrying way.

Hitler allowed correcting, to a certain point, the illusion of separation.

All suffering comes from the belief in an illusion. The truth does you free. All to your around and all that they do is a gift for you that it allows to know you yourself and to redefine you to yourself. You are in your world's point of cause.

Once you understand like a so bad person like Hitler it can be a gift and as a person like him is caused by the world that you hedge it in with way such of that the world may experience his mind and beliefs, you will understand a great secret toward the riches. When you understand than

Hitler acted like worldly mirror, a set of little pieces of indifference, belief in superiority and separation out of every person in the world of that one then, you will understand like you you can be a set concentrated of riches for the world. In other words, not fear dreaming at large, to dream of big riches and of believing that you can have a great riches. The world will take care that that way I succeed and in fact makes somebody have those big dreams insofar as you want to have them. You have all of the support, be as fuere the road that you choose to take.

If constants allow to themselves them besides things, an individual ó society they will catch up with the riches and happiness insofar as they understand and the law of cause and effect of the correct way be applicable .

What you have now is a good introduction to the principal law of the universe. As you read the following chapters, you will understand this law much more clearly, specially the chapters on Conditions, success and Carnal Cuántica. Now that you are put together with the knowledge of the law that you do to work to the universe, let's give what really they are the conditions to a look ó circumstances and success. You are about to enter one of the more beautiful tolerant and hopeful journeys that occasionally to time you had undertaken

Conditions ó Circumstances: They are fantastic illusions
Once I had a enough complex dream. You were in an idiom that it was, let's say symbolic, but the best way to describe it would be non-verbal.

One went from those dreams that one has when he is about to wake up, half a slept and I mediate alert. I became aware of what was happening .

An orange wall appeared in front of me and I began to pay attention.

From now on, a writing that symbols were seeming to be appeared but that I somehow to read it during the dream, the words began to appear and said something like " The Influence of the circumstances does not exist . He be believed " (And after the discussion kept on during plenty of while).

A voice of knowledge kept on reading in an idiom similar to a vibration while I looked and understood . The message had around five sentences of an all-round and sensible knowledge. You ended up in seconds and rapidly I got up and I took out my notebook to try from writing the exact words again. But the moment that I began to lose hope to find a lapicera, I lost exact words. One way or another, his composition was not linear like the sentences of this book.

However I, of course, to retain the understanding although I lost exact words. The essence of that message is in this chapter. It is an easier little to understand it in the event that you understood the real nature of the time of the quantum physics and. Once you understand the time fully you will understand that it is a scientific illusion. Einstein and another brilliant scientists have shown us as the space-time operates in reality.

We checked like everything, the past, the present and the future, all exist in the Ahora's eternal moment. But being participating children within this moment, we experienced time's sensation as we went by next to another participants in the continuum.

Just as we see in the Carnal Cuántica, enough is all the quantum soup literally those things and options that can exist and exist at the same time. In other words, all that you may imagine, and more, you right now exist and you exist at this moment of the now.

Everything! Then, one I rich and one I not rich they exist simultaneously of you, but you are experiencing only one of them, are conciente, I wake up and giving you an account of one of them.

Very well, he is time to put on thinker's his cap. Here we go : If everything exists simultaneously, in this precise instant, you can see like the conditions do they not exist like real properties of the universe? If you unlace all of the possible they exist, How can there be a condition in opposing view of than true unlace exist, when in reality they right now exist ?! Look, a moment does only, you read last sentence. You already read it. ¿As you can then to have joined condition that you have not read it when in reality you right now have read it? It is not possible. Then, because everything is possible, everything is conceivable, everything right now exists, is not possible that there be a condition to not to exist, in opposing view of his right now existence. Your rich being right now the scientist exists and spiritually. All that you need to do is to outmode your conscience toward that part of the Being that is delicious. Ó does not have conditions circumstances in order that that part exist ; Nothing can impede you existing because you right now exist, but your you can create others unlace that they seem circumstances. For example, something else that exists but you would be able not to as living it you are not to be rich materially, your to postpone ó sensing that you are out of the adequate moment, your going to the casino and getting in the habit of the game, being in a poor country, no having education, etc.
All these are independent existences that oppose the existence of the riches. But because they exist you do not signify than the rich existence not exist ó exist in terms of another existences.

The people commit an error when saying good, if I would have been born in a rich family, ó at a good country, ó with a determined Mr. ó knowledge, ó if I would have gone into in such ó as university, ó if such ó as sew it would have followed then you would be rich They Are Supposing That the riches have a condition but they name like condition what in reality it is another independent existence to. In other words, you can get to the riches with ó without having that to happen for grasped her named condition. You need not those conditions to be rich. When you believe in the conditions ó circumstances then there they will appear to you. There are uncountable roads toward the riches and conditions are only one of those roads – it is your decision to desire ó not to pass through the conditions. Inclusive the time, that would seem to be like a condition, is it not. The people think that it is needed long time and age to make a fortune and naturally happens to them that way but you do not have because being that way.

All that is conceivable exists ; You can understand this providing that you understood the nature of the time of them, as many as and of the spirit. Conditions are not sentences if such thing then such another one that they separate having with not having . They are another outcome of an infinite number simply of unlace . They do not exist like conditions; They exist like another possible existence. You do not have to pass through the conditions but if you believe in them you create them and then, of course, you are going to live.
What's important that is necessary to know is than, still scientifically, the conditions do not exist as if they be sentences.

If such thing then such something else, like requests you have to pass which, ó like traps in the ones that you have been born in and you must remain, ó like things than of the ones that you can escape of using certain actions and obtaining certain help that is out of you. The influence of the circumstances does not exist . We have designed the influence of the circumstances all by ourselves. What you call a condition it is not a condition another one of the infinite is possible existences simply and does not exclude you to have another existence.

Keep on reading and you will understand them and next you will surpass them.

The influence of the circumstances does not exist

The circumstances outside affect you only inasmuch as you allow him to him.

Ambient external midway and a person's circumstances are always related to the interior statuses and the person's thought.

You are through the environment and the conditions that a person experiences and discover your thoughts and statuses. This is made possible for the fact that the universe, without error, modifies the environment and the conditions that a person has as from his thoughts and statuses of being. A person always is in the perfect surroundings to see oneself and experimenting, in order to change and to grow. A person can direct and expediting his growth in the riches in everything and them besides areas acknowledging to and utilizing this perfect system.

Think out of the square. The square is the box that was created for your last conditionings, experiences, teachers, news, environment and suchlike things. It is not a real box; You exist only upon the mind, your mind and the minds of those that they surround you. This is all that maintains the square.

You are not real by itself; Needs you to you and to everybody to survive.

You can come out of the square thinking out of him. This you have listened to it many times but now finally you can do it all the times. You achieve it brushing aside all of the constructions that you have in the mind bring near of how things are done . For example, some people that never had an university education believe that to for to be rich they have to have an university title. They told them that and they think that they have observed it for which they maintain that it is truth. This limitation can be eliminated simply with removing the construction from his pansies and his belief.

Many have done it and had an impressive success. In fact, Bill Gates of the company the university left Microsoft voluntarily and she never obtained the title. Million people around the world have achieved spectacular successes without having an university title. This does not mean that you not have to go to university. The universities have a very important role but if you feel than in your square, in your box, the idea that you did not go to university and that at present you can not assist is, simply

You abandon the construction and the belief in that and the square decreases itself. That is what it means considering out of the square.

Can do it in any occupation from developing of a product, working in finance, in a new idea of business ó anything. You do it simply rendering an account to you, having the intention to do it and abandoning conciente constructions. Ask : ¿How do you construct a house ó do you manufacture a car that they be radically new? You do it abandoning all of the previous constructions of what they told you bring near of how ó builds for itself a house a car is manufactured. Abandon them and start up zero to sheet without no you would owe and you would not owe and next you abandon the tool. Quit that you receive everything. The key is the inspiration without the restrictions of the you would owe you would not owe . It is something very intentional, wild and very fruitful. Ask Einstein.

Do that all the conditions be useful for you since for that matter they are . They are a field of experience created entirely for your joy, auto-discovery and learning. This is due to that they are die-formed as from your previous thoughts, words, actions and statuses of being.

Confront all of the financial and another- type difficulties calmly.

Before they happen, they right now have been solved . Before you ask it for right now you have been given it. Simply receive .

Ó does not have coincidences, accidents, chances luck in the universe.

The universe works under perfect laws that they never fail, not even once only. The Source, God it works to perfection. Everything works perfectly by law. Things seem coincidences and accidents for those that they do not see the truth behind theirs.

The nature works with precision and I save total of effort, with an infinite organization, with incalculable roads, without a struggle. You do not need to understand as the riches is created. Simply maintain the loud vision, think, speak and act in accordance with your vision and automatically, by coincidence it is going to work. Be unable to bear with what happens on the way over since part of the incalculable roads that nature utilizes is for bringing you the that you have visualized . Simply maintain the part of the deal that concerns you backing up your vision with firmness and thinking, talking and acting on the basis of your vision. Himself impartial, detached; This permits that the creativity of nature operate . The detachment means not preferring different nothing so that it is happening at this moment and to have the freedom to elect a different future.

What happens at this moment is the perfect manifestation of your intentions, thoughts, words and last actions. The act to prefer an alternative Now, it leads to the delay of the achievement of your goals. Preference is to want and wanting perpetuates the status to want.
For example, if your desire and intention is give it becoming multimillionaire and you right now have obeyed the part that concerns you aligning your thoughts, words and actions with your vision, you must admit that you do not know as the best and optimal road is for your goal being enough.

You can not predict the exact events, daily, that they will lead to your longed-for result. The Source can make it easily. Your Inner Being might also to do it and taking of the best possible way. Leave her to work with your magic no resisting in a minute of the Now That you are showing you. Do your part and leave her that she do her part . It is the most efficient, effective and amusing way.

You always can be happy and relaxed because you know that it is what comes at the end.

Calm is to can, the calm puts you in accord with yourself and with nature. You put you on control knob of your thoughts permitting that you have the correct thought. It is your testament to the fact that you are not circumstances and that you are not the circumstances inferiorly. Calm is confidence. Calm is your true nature, perfect equilibrium and perfect peace. Always say, I am calm.

Calm is not the suppression. Calm is clear and open; Do not hide neither suppress nothing internally in order to show calm. It is pure and natural. The calm, represent and she can be developed by means of knowledge, interiorizing, practicing and living the laws of the universe you see like in this book, the laws of life, such.

The suffering suggests that there is an error in thinking. A clue is for finding and to change the error. Very often, in the moments of the more deep suffering you find the bigger opportunity to find a new truth. But suffering is not necessary . In fact, the more illuminations could have eliminated the suffering completely.

Suffering is only a means of communication from the Being toward your personality, your mind and of return to the Being to put it of an abridged way,. It is utilized only if not there are another options. The people what else stand against his Ser's subtlest clues music them what else they suffer . Those that are more attentive and tuned with the Being through things such like the active intuition, are able to work with life deliberately instead of that life operate with them.

The conditions do not exist . The influence of the circumstances is an illusion. Conditions are fabricated to create an environment that correspond

With your thoughts. In other words, the influence of the circumstances is an illusion that obeys the thought to satisfy and to manifest the thought in the experience of that thought. The conditions turn the thought from symbols into the mind to real experiences.

The fact of the matter is that the universe contains all that may exist and everything it is contained in the Ahora's eternal moment, Aquí. If you think and you believe that you are poor, they will create conditions in your around in order to fulfill that belief and those thoughts. Then, the declaration of I Can Not Buy It because I do not have the money it is false. What in reality happens is that you believe in insufficiency and the world around yours you get comfortable in the time to present you needs of things that you can not buy. On purpose and for the same reason, the need also is an illusion. ¿How can you need what right now you have ?

You right now have everything because everything right now has come into being . (before you ask it for right now you have been given it, like Jesús and many other teachers have taught us and he is what's same that now the Carnal Cuántica says us)

The conditions do not exist . The influence of the circumstances is an illusion. You are the that you cause conditions but seem as if they be for imposed to you by external forces. This is one of the most liberating knowledge. Understand it in depth. Live for him and life will become magic. Simply attempt it.

When you fight against the circumstances, you are fighting in vain against the effects and preserving the cause. Let's say, for example, that your circumstances are such than to you broken time. If you act from the perspective of the bankruptcy (reducing costs, acting brusquely, being a miser, bitter, fearful and jealous) in an attempt to avoid you to be still but fraction, What is what really you are doing ? ¿You can begin to see as you perpetuate and do you energize the condition to be broken?

By means of believing and thinking consistently about being broken, these creating the conditions of bankruptcy utilizing to can of belief and the thought. Remember, the universe always gives you to that your you think more often and believing it. You are to your service. The way to put in order the problem to be broken is to assume a status of interior riches and thinking, to talk and to act from that perspective.

The riches are first a status to be that one experiments and upside-down no.

The riches do not create due to certain conditions rather certain conditions themselves they believe themselves due to the riches. The abundance does not create for oneself because of certain conditions but they create conditions for oneself due to abundance. The influence of the circumstances does not exist . The following declaration is false: A person is poor due to certain conditions that are surrounded. The following declaration is true: The conditions of poverty surround a person because his Ser and thoughts are their own of a conscience of poverty. The status to be creates conditions. They create the people's majority that you are upside-down. Those that see this in a clear way verify that their conditions transmute magically giving them lucky breaks and coincidences.

You stop being in the thought that you are your conditions. Say : "I am not feign it the fact that I would like to have, I am not my work, is missing of riches I am not conditions "

Why do the things recapitulate? For example, the people that always fails in any business that she undertake are . This is the reason: You keep on attracting the same circumstances and situations time after time until you decide to have fun to new, changing your employers of thought and changing your being to new to turn it into a new improved being

You avoid judging the things like correct incorrect ó. Things are things simply. His classification comes from the election of the observer that classifies them like good bad ó, correct or incorrect, amusing bored ó. The moment that you pass judgement to the things, you pass judgement to yourself.

Also you block the occult gift that an event brings you. The Taoístas have a wonderful form to explain it using the farmer's history.

Once upon a time a farmer to who had escaped to him the horse and his neighbor Siento told him Very Much that you had happened to something so terrible. The farmer answered you: "You not feel uncomfortable because who what bad ó is good knows ".

Good, on the following day the horse that had escaped returned with savage horses's tropilla you had become friends with. The neighbor told you:
" I Congratulate You, than good luck! The farmer answered back : "Not congratulate me because who what bad ó is good knows ". On the following day the farmer's son tried to mount the savage horses's one and he broke a leg. Again the neighbor came and told the farmer Lament so much so that you had happened to a so terrible thing. The farmer answered back : "Not deplore him because who what bad ó is good knows ". On the following day the soldiers got to the farm to compel people to serve in a military force to take them to the army but the farmer's son was excepted because you had the broken leg.

This is a simple story but illustrate the miracles that you make The Source, of the most unpredictable manners and apparently disconnected in order to that everything work to perfection. For those that they know and they work with the laws of the universe, that apply the law of cause and effect, that safe goals, purposes and visions have the magic it works perfectly, bringing synchronized events to them, coincidences and many others spins and turns that conduce to the desired ending.

Therefore, you avoid passing judgement to the things and to the people. The one thing that you achieve is to rein in you and getting damaged since you do not know the chain of events that they have glided . In like manner, you become what you judge and that that you condemn turns you to affect you. You condemn for law of cause and effect, when you judge ó, you put yourself in a situation to stand for trial and convict. The part that you are in for is to have an obvious vision and a certainty in all the themes that relate with your life, that the internal functioning is of how your goals are going to come true. The daily events, good bad ó will take care of you in the meantime do not interfere with the process.

Your intention works like a magnet that all that is necessary to manifest it in the world physical attracts . This is an example of how the intention works : You think about a new idea, you have a new desire. You have the intention that desire become manifest of. This, then, begins to attract everything sew her that they require to attain your desire. It is something magic – you have certain dreams, you know a determined person, you suffer through determined conditions, you acquire determined abilities and everything apparently seems to happen like coincidence. As you advance this process, parts of your desire continue being shown at your reality even than the total of the same you become manifest completely. You rely on the system. Do not battle against the things that you attract your intention since your co's perfect designs are creation.

You watch with humor difficulties – it is the first step toward becoming unattached of them.

Cause sometimes does the people hit rock bottom? ¿Why does reason be necessary to times the people get devastated financially, emotionally ó otherwise? The answer, lie in the use of suffering. We are not designed to suffer. If we listen to our spirit much more, if we recover the capability to be much more intuitive and listen to intuition, our suffering would decrease radically. When we refused to listening to wisdom lifted of our soul, we suffered errors in thinking in the world physical as a corrective measure. ¿Occasionally you asked yourself because reason were many people that made a fortune outrageously occasionally bankrupt? The classical history of of the rags to the riches.

Remember, it is not necessary for this to happen but when it happens and the phony individual takes the responsibility and learns from it, she becomes very rich.

When the person hits rock bottom he is once everything was stripped out of. Retreat to him all false psychological defenses that they impede that a person accept her Loftiest Truth. When you give the person you identify and it becomes disconnected of these falsenesses and you accept the Truth, rise again very high.

In relation to the riches, one of the Truths is that we are abundant by nature. There are many Truths on the riches and the happiness and they treat a great their deals in this book. It is not necessary for you to hit rock bottom and backbands to identify with the Truth. The suffering only comes when the person refuses flatly to listening to more subtle signs herself.

Those signs can come from ó of sources inside such like books, someone else, TV, films, etc. The Source communicates of infinite manners. We are the ones that we did not pay attention to him neither we listened . We are the ones that we were ignorant of that what we felt is true.

Another reason for suffering is to know what suffering is not and like coming to the no suffering. For example, in order that you may know what the red color is, you have to know what is not red. You do not have because knowing this for experience and personally but definitively you have to know.

Imagine a robot that never experienced the status of happiness ó sadness.

It may be that the robot be very intelligent, very well programmed in the knowledge of happiness and sadness. The programmer is able to explain as much as what the happiness is be possible but the robot is going to know the very happiness conceptually.
This is not the true knowledge; It is empty. Only experiencing the happiness by yourself you can know the happiness and you need to live for it although he be for a moment, the opposite of the happiness that the sadness is . There are something that you need to live on them first coat and others that only you need to know conceptually. Sometimes, the things that you need to know of first coat include what suffering names itself but in essence, they are tools that allow you to enjoy the opposites that you look for .

Very much auto-tax comes from the pain imposed. It is the bitter potion which the interior doctor remedies you the sick Being by means of. Therefore, you trust the doctor and drink the remedy in silence and tranquility right now than the doctor's hand, although weighing and last, she goes upon the Invisible Khalil Gibran's new hand

The psychological rule says than when you do not get a situation to conscience, happens out like destiny. This is, when an individual remains not divided and conciente of his own contradictions does not turn around, the world of necessity intervenes in the conflict and it is torn in opposed halves – Carl Jung -

Leaving everything them besides constant things, an individual or society will catch up with the riches and happiness insofar as they understand and use the illusion of the influence of the circumstances of the correct way.

Refreshing knowledge proves to be that we are not at the mercy of circumstances and conditions at random, knowing that we created them and the fact that the gifts that serve to serve our growth are beautiful. ¡You are liberating!

Now we will examine an aspect of the influence of the circumstances that has been extremely distorted in the society. Rumor has it that one can have success ó failing . This is the beautiful truth and secret: You always have success

Success:
You never can fail.

In relation to the riches, there are a condition that had been more distorted than the success and the failure. A great many people think that one cans, ó well having success ó well failing . Good, the failure is another illusion and the one thing that is is success. A so deep truth is this that you should do the effort to understand her.

All life is a fraught success of successive moments

The failure, when you utilize it without fear like process of learning he is what the weakness eliminates and construct the fortress in thought and character. This new fortress in thought and character is essential and criticize for the future success that you search . "The failure " a moment is of really success, a moment of success in its own right. One learns like to have success through the failure and as the taste is of the final triumph. ¿How would you learn to delight the sweet taste of triumph else you know as the taste is of the failure? ¿How would you know how to come to the triumph of destination without the tools designed to arrive?

The cause but common the lack is of an obvious and focused goal of the failure and is not enough it of visualization. The universe, Life, literally you lack material to work and then you do nothing . The imagery of the mind are the life, expressed. There is no expression without imagery.

Often, in your darkest hour, lie the bigger opportunity to be all that you can be. Your worse moments are, frequently, your older liberators, your loftiest teachers. Do not battle neither resist them when they touch you.

In his place, examine them to find the lessons that they contain and the liberation that they bear . Suffering is always an indication of a wrong thought. You are not designed to suffer. You were designed for enjoying life.

Profits and losses are two faces of the same coin. Through the loss, new desires things. You know the sweetness of the profit through the loss. There would be no profit without loss. What causes suffering and retarded growth is the fight against the loss and the preference of the profit on the loss. The approval comes from the loss and the profit as much as a gift like a fuel for your growth that is going to propel you a bigger height and. At the end you will see that that loss was not a loss really. At that point, when you recognize the profits that you got from the loss, you will see that the loss resulted in a blessing and that the loss does not exist . Each loss is 1 usually) The result of an error of thought, in which case you have opportunity to correct and to have enormous profits or 2) once a new selected opportunity Was Not Owed to your thought but for your I Superior designed to take higher, a possibility to discover a news and more lofty truth.

Utilize all the conditions to have a good time and to be built; For that matter they are . Even the ones that are negative are useful . If, for example, you come to grips with a situation you are in under oppressive people, be assured that you choose to be in a liberating, love status yourself, an interior status of love (what's opposed to oppression). Think, talk and you perform on the freedom, the love to yourself and love.

Show them fine gesture and pardon instead of being oppressive when you have the opportunity . The way you can leave in of your negative circumstances is that one and besides visualizing, with belief, the way you wish to create from now on your world in in detail. When you become a person not oppressive that herself and the other ones are loved, really, internally, you will see yourself to yourself coming out of that oppressive position. The negative situation will have done its work when redressing you with your voluntary participation. Remember provided that you elect certain level, you the circumstances in which you are implicated, that he be ó in not than be conciente of it.

When you look in yours to find the error in the thought that causes your suffering, look always humbly in order that your ego not come in the way . Look at with sincerity to find sincerely. Ó watches without self-pity pity to the sword call her sword. Remember, it is a private exercise; Nobody will know neither will do the ridicule with you so that you can feel yourself free to be precise and honest with yourself.

You have been the people's majority programmed to fear the failure. They are beaten to avoid the failure, ó not even they try that way they can not fail. The failure, however, is an illusion. Begin to see it like an illusion. The failure, suffering are one component essential of success. It is what you help you to correct the incorrect thought if you focus it with an attitude of learning. Through the failure, you learn like to have success. Trial-and-error, refined your thought and you aim it more and more close at success. This is that way only if not you are beaten.

Through the failure, you manage to know the success and I eat to arrive. ¿How else would you know the taste of the success without elks to know what is not his taste? ¿And how would you get there unknowingly like? You think about this. The failure is an integral part of success. The failure is a moment of success that leads to the principal of successes in reality. The failure is not what's opposed one belonging to success, a separate entity is success.

The failure is success, it is all the same but in the specter's different extremes, the specter of the achievement. It is as you warm up and cold: Extremes of the specter of the temperature in a thermometer are different. Success and the failure are different vibrations of the same thing.

The failure is not failure as such. Only it is failure when you accept it like the end. But, if you accept it like a blessed part of the process, one part that helps to have you more success and namely I eat is the taste of this success that comes, then never you can fail, never. The failure is an illusion. You stop to fear him; Love it for the gifts that you bring you.

Life is a set of experiences. The challenges are a part of experiences. Use the challenges to improve you and to enjoy the reward and the triumphant experience that you follow to each challenge.

Each attempt is a success that leads to the final result that you look for, the great success. Look at it that way.

Your survival is guaranteed. You do not have to obey requirements for your dignity and your life.

Quitting you fix everything them besides things an individual ó society they will catch up with the riches and happiness insofar as they foment and applaud all of the moments, events and attempts, seeing them like successive moments of success.

Now that we have defined the success and taken the top off the illusion of the failure, it is moment from examining the object of the quest of success. In other words, that the people want . There is another secret here: Desire, but never want

Do not want :
Desire, but never ever want .
A good reason to watch your words, pansies, stock and states exists . By law, each status and thought, act in accurate form on the universe. Each word bears thousands of years and instructions of how it will be manifested. For example, the word jumping invokes specific imagery in the mind of anyone along with the specific instructions to execute it. And, the universe, that has to help in that jump, act in consequence (laws of physics, coordination between mind and spirit, etc.) Enclosure to read this book, the words invoke in you certain things, some which you can feel of at this moment.

Some people will feel enthusiastic when learning the words that they will allow them to here to do a great difference in their lives and than knowing right now you are beginning to produce invisible changes. Some people right now know it in the meantime they read this.

The more important word to beware of her is in relation to the riches I want and all his equivalents. Wanting communicates you to you and to the universe that it is in that you do not have (first error) and that you are in the status to not to have it but that you would wish to have it (second error). The problem is aggravated by the fact that the to want meets in a perpetual status. By itself you do not have purpose. Consider it.

You never can obtain what you want . Never.
Seem as if the people get what you want but in reality he is not that way. What happens in reality is that gradually they move around from the status to love another states and then they get what initially they were wanting . But in the meantime be in the status to want, they can not get what they want . The way is this in than works the illusion that a person gets what you want : Remember last time that you wanted to eat and you got what you were wanting (you ate something).

Very well, you wanted something to eat. A status to want is that one. But what happened from now on looks . You began going to get something to eat.

In reality that's where you moved around of the status to love the state to get hold of that it is a purpose. Next you moved around toward the present tense to have and finally you appear having what you were wanting . As you can see, you never obtained what you were wanting when you were in the status to want. You had to change status.

This irresponsible change of the status to love another state is something than everybody all the time ago but solo for the little things.

What would happen if did he be something big, something that never before you did ? ¿Would you get it if that way you want it? Unlike food, you would be more difficult for you to change unconsciously to move around of the status to want granted that you never have gone by over there previously. Definitely you see yourself wanting twenty dollars, it would be easy to displace you unconsciously from the status to love the state to get because you have done it time after time in the past. But ...

What would happen if you want one million dollars and never before did you have over twenty thousand dollars in your life? ¿Would you move around unconsciously from wanting the million dollars to have it? Probably not. The solution is the following: ¡Never want !

You never can get what you want . Wanting something with a lot of intensity is worse . In your thoughts, words, statuses and sentiments you replace wanting with desiring or longing for.

It does not have to mean that you do not have something unlike wanting, desiring necessarily. It is a very subtle difference and some people can say to want and to desire they mean the same thing but a world of difference is . Some synonyms dictionaries might even to say than the word saying and the word wanting they can interchange themselves but that is only for some linguistic purposes. Remember, your thoughts are executed by the universe with perfection and precision. The way in which the system is designed is that one. Wanting is executed with precision and represents a perpetual status to not to have.

Desire is not a perpetual status to not to have; In fact, necessarily it does not have to mean that you do not have what you desire . To think that thousands of million people are kept away of what they want for a so simple and so small difference it is saddening and amusing. Everything is reduced to precise execution of the universe.

Precisely, the word does not stand alone I want the one that should be avoided.

It is the status. It is useless avoiding the word I want without avoiding to be in the status of I want, it's no use. The language is an used symbol to represent things like a status. I want the word it is a symbol that represents the status to want. The state that you should avoid in low gear is instance therefore. The symbol, the word same, also it is necessary to avoid her not to invoke the status. Please, desire, but never want

There are some definitions of the dictionary for the word here to want: Being without; Lacking . To be destitute needy ó. A defect of character; A fault.

Being away; Being deficient empty ó; No being sufficient; Being lost for words; Lacking .

This is what you communicate the universe in order that you bring something when you want it to you. The universe brings that to you simply – absence and deficiency. No one of those definitions they include in the word desire. There are some definitions of the dictionary for the word here to desire: Making a wish; Requesting .

The natural longing that is stimulated by joy ó the thought of a good thing and effort of his continuation or possession encourages ó to the action; An avid desire to obtain ó to have a good time.

No needing a particular result in this moment of the now you free the subconscious from all of the thoughts that you can not obtain a result about in particular. This, in turn, paves the way for the individual proven to be that he was proposed conciente. This is one of the benefits to have the intention with detachment. You have the intention of a future outcome, you have certainty of it, but you are once what you pass at present was separated from. For example, let's suppose that you mean to be worth millions, but as of the present moment, the events move themselves so that they suggest that you are not going toward your goal. You will do the best progress if you are taken off in the present moment what the fact that you accept it instead of fighting against him, frustrating you and losing hopes means . Independently of the detachment, maintain the certainty of the future outcome on becoming millionaire. Learn how to allow the co existence of the intention, certainty and detachment in your life and life will rapidly become happy and confidence will grow . As you can see you can not fail to make a fortune if your goals are clear, concentrated and you are certain, faith and you believe that it is impossible to fail.

The failure would mean breaking the unbreakable laws of the universe. Therefore the certainty allows to relax you knowing that the riches are on their way, independently of what the present-day moment may seem. The lack of detachment is equal to resistance and what you resist persists .

Eliminate all the forms to want. This includes the wailing about the past, desiring that things would have been different, waiting than, desiring than, wanting, getting worried and pushing ó toward the past to the conscience toward the future. In other words, not to you to aferres the last moment, you not wish to be in the following moment and you understand the moment of the Ahora in complete form to welcome all the Mr.s that you bring you in his place. The fastest the way to create a great future is simply to have the intention, to liberate her and going back to have a good time here and now. Wanting says the universe that he has to create the conditions that they maintain you in wanting, doing impossibility that you have what you want to the least that you get out of the status to want and passes to another status. You are very subtle but important form to see the life.

Never want nothing. Wanting does that the universe give the conditions to create a perpetual status to want to you. Have passion, desire, have intentions but do not want .

Eliminate the word I want of the language and the status to want in your thought and sentiment. Replace her with desiring and desiring, being desirable and being desirable. Wanting creates the conditions of I perpetuate wanting . You never get what you want .

In the event that you meet thinking that you do not have something ó that you are not something then these wanting . To want is a status to be that you profess not to have, the scarcity. A word is not simply of the spoken language.

If everything quit them besides constant things, an individual ó the riches and happiness insofar as they eliminate the word wanting of his language and state to be will attain society.

If you see yourself to yourself battling against wanting, you go back to read the chapters on Carnal Cuántica, Abundance, One and Being and understand really and just as the teachers during thousands of years taught it than have it everything and before you ask it for right now you have been given it. Logic, scientist and spiritually there is a reason to want absolutely.

Of course, a reason to desire is, but no one to want. To want is a belief to not to have. You right now have everything. ¿Why would you have to want to believe that you do not have it?

Now that we already clarified everything that, let's look at some big and interesting things. We began with your Purpose, that only yourself form that nobody else can duplicate. Let's continue

Purpose:
What you are here.
As it your purpose? Your purpose is to separate you from the goals. ¿Which is your purpose and because you are here? Only knowing and declaring your purpose and maintaining it loudly you can move every day quickly in the correct address and to pass her very well on the way over.

The correct question is, Why did you choose to come here to earth?

Where did you obtain your purpose? You consider this suggestion: You have free free will.

Where did this free will begin? Some people consider that you begin when you are born . Create that they did not have election of being born ó no being born, but once they are here, they have free free will to elect.

Other ones believe that your free will is eternal and that you begin still before you be born . This is not a strange idea. The soul is eternal. Your eyes say you that life begins when one is born, but something deeper you tell you that this is able not to be the real beginning. The purpose I destine ó that is you, your Ser ó nourishing they chose to serve to do to earth, an election based in his set of attributes and desires. Besides circumstances and position of the place where were born they are perfect to accumulate the necessary tools to fulfill your elected purpose if he is than be conciente of it when going by life. He is your singleness of purpose that's why you make somebody feel your so good when you find it. That's why you feel so much joy and I enjoy the to do it. That's why you elected it you make time so much. That is what you came to do here.

Unfortunately, a great many people fail to satisfy their purpose largely due to the way in which the society is structured and education. Still that way easily you can fulfill your purpose if that way you elect it.

You find your purpose passing a certain time in silence thinking that he is what you make you feel well first, stops than sew you have a lot of passion. Do not think about the description of a job or race.

You brush aside all those labels the society has taught you how to in to believe. They are the number one reason which the people do not find his purpose for. Ask yourself to yourself: "¿What is it what else do you gladden me? It would be able to be to be with the butterflies, to pilot around the world doing business, cooking, talking with the people ó anything else. Once you do so, visualize it, have the intention and put goals toward the point where you are at work, like race, in your discovered purpose for you.

For example, let's imagine that your purpose be to study butterflies but that you work unhappily on a job that does not have to do with the butterflies. Do not lose hope . Begin to get books and to look for people at the field of study of the butterflies. Everything finds out that be able to . Next you begin to put you goals and elections that at the end they lead you to a job that be related to the butterflies. Do not worry about money and besides smallnesses that begin to appear to you alone they are going to get solved in the meantime and as soon as you not get worried . You are going to be very happy and successful once you be working on your destination ó purpose, a destination than yourself you elected before having been born in earth. You car satisfaction will rise and you will be doing an optimal contribution to the world in general.

END

www.ingramcontent.com/pod-product-compliance
Lightning Source LLC
Chambersburg PA
CBHW081055170526
45166CB00006B/2075